DEPAR' CW00322471 MENT

O×R ✓

Air Pollution & Tree Health in the United Kingdom

London: HMSO

© Crown copyright 1993
Applications for reproduction should be made to HMSO
First published 1993

ISBN 011 752636 3

44 8347

Oxfordshire County Libraries

Date	14.10.93
Class No	Y/0/634.9 ENV
C	DAN

Cover photograph: Severe ash dieback in a tree now isolated in an arable field though located on the line of an old hedgerow.

Executive Summary

1. This report has been produced in response to a conclusion of the first report of TERG published in 1988 which stated that there was no direct evidence of pollution-related forest decline in the UK but recognised that the pollution climates of some areas could contribute to tree stress.

2. In this second report new data from tree health surveys, controlled experiments and pollution monitoring networks have been reviewed with the specific aim of assessing the relative importance of atmospheric pollution as an influence on tree health in the UK.

3. Improved monitoring networks and continued research have led to an increased understanding of the various pollution climates in the UK. The importance of cloud water deposition and altitude enhancement of wet deposition on pollution inputs to upland forests has been confirmed. The significance of ammonia as a component of pollution climates and nitrogen critical loads has also emerged.

4. In the UK, the majority of commercial forests are located in the north and west of the country, often at considerable altitudes. Wet deposition of pollutants is large in these areas and concentrations of major ions in hill cloud have been shown to be very large relative to those in rain. Consequently, in these upland areas, tree foliage is exposed to very large concentrations of pollutants such as sulphate, nitrate, ammonium and acidity.

5. Because of their morphological characteristics, forests receive greater inputs of pollutants such as nitric acid, ammonia, sulphur dioxide and ions in cloud droplets than nearby short vegetation.

6. In central and southern England, inputs of sulphur and nitrogen to trees are dominated by dry deposition processes. Deposition of gaseous ammonia represents the largest input of nitrogen to unfertilised vegetation in these areas. There is increasing evidence that the presence of gaseous ammonia may increase sulphur dioxide (SO_2) deposition and vice versa.

7. In urban areas of central and southern England, amenity trees are exposed to the largest concentrations of SO_2 and nitrogen dioxide (NO_2) experienced in the UK. In rural areas, concentrations of SO_2 and NO_2 are lower but forest and amenity trees are more likely to be exposed to frequent ozone (O_3) episodes.

8. Concentrations of potentially phytotoxic ozone (> 60 ppb (parts in 10^9 by volume)) occur as episodes throughout the UK. In the south of England, O_3 episodes are far more frequent than in the north of Britain. In warm years many hours of potentially phytotoxic concentrations of O_3 can be expected.

9. In surveys of tree health, progress has been made in methods of assessing crown condition. The most recent UNECE (United Nations Economic Commission for Europe) survey of tree health in 1991 shows the condition of UK trees to be generally poorer than in other European countries. It is thus essential that development of future survey programmes should include increased liaison and co-operation with other European countries so that efforts to understand forest decline can be more effectively pooled and survey methods standardised to improve international comparisons. In general, UK survey results do not indicate a widespread decline in crown condition, but at some southern sites there is evidence of recent deterioration in the crown condition of beech.

10. There are many damaging agents other than atmospheric pollutants that may affect trees. These include adverse weather such as high winds and unseasonal frosts. Soil factors can also be damaging. In lowland and upland

areas, soil water availability can limit growth. Nutrient deficiencies can restrict growth, often inducing distinctive foliar symptoms, and excesses of certain heavy metals can be toxic, causing poor or malformed growth. Conspicuous damage symptoms in crowns of mature trees are frequently caused by biotic agents such as fungi and insects. With such a variety of damaging agents, it is not unusual for trees to show some evidence of past or recent damage.

11. Considerable experimental evidence of affects of atmospheric pollutants on trees and soils has been accumulated since the first TERG report:

    i) Filtration experiments have shown that exclusion of gaseous pollutants improved the growth of beech and Norway spruce at a site in southern England but not at sites in Derbyshire or in Scotland. The evidence available suggests that ozone is the major pollutant responsible for the observed effects in southern England.

    ii) The potentially damaging effect of ozone on tree growth has been confirmed in fumigation experiments. Exposure to realistic concentrations of ozone has reduced the growth of beech and Sitka spruce and there is also evidence that effects over several seasons of exposure are cumulative.

    iii) There is little evidence that current rural concentrations of sulphur dioxide and nitrogen dioxide are directly reducing growth of major tree species in the UK.

    iv) Experimental applications of acid mists (at the largest concentrations found in the uplands) have reduced the growth of Sitka spruce. Reductions in the stem volume of 18-year-old trees occurred in the absence of visible injury symptoms.

12. Realistic concentrations of gaseous pollutants and acid deposition can cause subtle changes in the morphology, physiology and biochemistry of trees without obviously reducing growth. These changes may affect the sensitivity of trees to environmental factors which are known to be important damaging agents in the UK. For example, ozone can increase the sensitivity of young trees to autumn frosts and winter stress, and impair control of water loss in conifers. Exposure to sulphur dioxide and nitrogen dioxide at concentrations found in the more polluted areas of the UK has been shown to increase the performance of certain pests such as the green spruce aphid.

13. It must be emphasised that much of the experimental evidence of effects of air pollutants on tree health is based on relatively short term experiments on seedlings and juvenile trees grown in chambers. Experiments on more mature trees are taking place at a number of sites in the UK but they are costly and difficult to undertake. Research is needed to determine the limitations of extrapolating from effects on young trees to effects on mature trees and ultimately forest stands.

14. There is increasing evidence that forest soils in parts of the UK have been acidified by deposition of pollutants. Such acidification may lead to various changes associated with soil acidification such as nutrient imbalance, metal toxicity, changes in populations of mycorrhizal fungi and reduced rates of organic matter decomposition, with consequent implications both for tree health and surface water acidification.

15. Acute air pollution damage to trees is rare in Britain but a number of specific cases of tree damage in which the involvement of air pollution was suspected are discussed. These highlight the difficulty of attributing specific

causes to incidents of tree damage, although in two of the most recent cases, in eastern England 1989 and North Yorkshire 1990, there is strong evidence of an air pollution contribution.

16. The review group conclude that:

    i) In some areas of the UK, it is likely that the pollution climate contributes to tree damage. This conclusion is based on a review of the latest experimental evidence and a survey of observed pollutant concentrations, especially those of ozone and acid mist.

    ii) Results of tree surveys are still inconclusive. The state of tree health in earlier surveys in the UK could not confidently be compared with tree health in other countries, nor could trends in health be discerned. However, the most recent UN-ECE survey of 1991 shows tree health in the UK to be generally poorer than in other European countries and shows that tree health has declined in recent years in the UK. Evidence of a recent deterioration in beech health in some parts of southern England is also a cause for concern.

    iii) Trees in the UK may be damaged by a wide variety of agents, such as wind, frost, drought and pests. Atmospheric pollutants can alter the sensitivity of trees to some of these agents and may also directly affect others. Consequently, an indirect influence of atmospheric pollutants cannot be discounted in some cases of tree damage.

    iv) Forest soils in some areas of the UK have been acidified by deposition of pollutants. Relative to short vegetation, forests receive greater inputs of sulphur and nitrogen to soils and catchments, with implications for freshwater ecology and water quality. These effects of afforestation need to be addressed in land use policy.

    v) Long term programmes of survey, monitoring and experimentation are needed to assess the impact of atmospheric pollution on tree health. Recent research and new monitoring techniques have substantially improved the knowledge of most aspects of the pollution climate in the UK. The recent recognition of the importance of ammonia merits further work to monitor its emissions and atmospheric concentrations.

    vi) Where instances of tree damage occur, multi-disciplinary analyses should be made. These analyses should draw as appropriate on the skill of specialists in fields such as pathology, soil science, pollution science, entomology, mycology, plant physiology and climatology.

Contents

List of Plates

List of Figures

List of Tables

1

Introduction

Background

Over the last decade, concern has been expressed in many countries of Europe and in North America over the general state of tree health. This concern has led to a number of national and international research and monitoring programmes in an attempt to better understand the problem and its causes. Initially it was suggested by some that a "new kind of forest decline" was occurring and that the cause of this was air pollution. Now it is realised that the situation is much more complicated. Air pollution is clearly a major factor in the serious problems that exist in some countries, but elsewhere other factors are important.

The United Kingdom Terrestrial Effects Review Group, in its first report published in September 1988, looked at some of the available evidence for tree decline and concluded at that time "there is as yet no direct proof of pollution-related forest decline in the UK, but some forests are subject to pollution climates which may be expected to cause stress. Isolated trees in hedgerows and urban areas may also be at risk. On the basis of the circumstantial evidence, it is recommended that surveys be maintained and more specific diagnostic tests for pollution damage be developed". The then Minister for Environment and Countryside, Lord Caithness, identified this area as one for further examination, bearing in mind the new scientific and survey data that had arisen since the publication of the report. In January 1989 he commissioned a sub-group of the Terrestrial Effects Review Group "to review new data from tree health surveys, controlled experiments and pollution monitoring networks in order to assess the relative importance of air pollution, other man made stresses and natural causes in observed declines in the health of UK trees".

What are the Issues?

In undertaking this commission, the sub-group decided that while the prime issue must be to determine the state of tree health in the UK, the secondary issue would be to decide whether any observed decline in tree health can be linked with atmospheric pollution or with other factors which affect tree health. As the task of this report is to assess the relative importance of atmospheric pollution on tree health rather than tree health *per se,* a thorough assessment of the pollution climate in the UK as it relates to tree health is provided as a background against which issues can be discussed.

Current concentration of pollutants in the atmosphere and rates of deposition are given in Chapter 2, where attention is drawn to the influence of topography on deposition and there is discussion of how geographic location and altitude affect the pollutant exposure of forests. Importance is also attached to the role of trees themselves in determining rates of pollutant deposition and in influencing pollutant inputs to soils and catchments.

Estimates of tree health on a large scale can be determined by means of carefully constructed field surveys and many have been carried out in Europe in the last decade in an attempt to assess the extent of 'forest decline'. Not all have been reliable and inconsistencies have frequently occurred. It is also difficult to make comparisons between different countries. The results of surveys of tree health in the UK are reviewed in Chapter 3. The criteria used to describe tree health in the various surveys are discussed and special emphasis is placed on methodology and comparability. The usefulness of surveys in identifying tree health at a local or national level is addressed, as is their ability to generate cause-effect hypotheses.

The Review Group recognises that many damaging agents other than atmospheric pollutants may affect trees and that more than one agent may affect a tree at any one time. The major causes of damage and death to trees, excluding atmospheric pollution, are described in Chapter 4. These include climatic factors such as drought, wind or frost, soil borne factors such as mineral deficiencies or toxicities and biotic agents, particularly fungi, insects and mammals.

Over the past 5 years, considerable research effort has been directed into experiments on the controlled exposure of trees to atmospheric pollutants. These experiments have been aimed at reproducing symptoms of tree decline observed in the field and at establishing the mechanisms of tree injury by pollutants. It has been important to recognise that tree response to air pollutants and other stresses, may vary according to the growth phase of the tree. The results of controlled exposure of trees to pollutants are reviewed in Chapter 5. Direct effects of various pollutants on the growth of a range of tree species are reported, and the effects of pollutants on tree sensitivity to environmental stresses such as frost and drought are also described. Air pollutants not only affect the tree canopy but also act on the soil both directly and indirectly via the tree canopy. Pollu-

tion-induced nutritional disorders arising largely from soil acidification have been extensively researched in Europe and the USA. The results of experimental studies as they relate to soils in the UK are also assessed in this chapter.

An important section of this report, Chapter 6, considers specific cases of tree damage where the involvement of air pollution was strongly suspected or claimed. A number of case studies illustrate the difficulty of attributing specific causes to incidents of tree damage although two recent cases appear to have a strong air pollution link.

In the final chapter, conclusions arising from the report are critically assessed and recommendations are made for future research and monitoring.

2

The Pollution Climate and Tree Health

Conclusions

● The majority of commercial forests in the UK are located in the north and west of Britain where wet deposition provides large inputs of SO_4^{2-}, NO_3^-, NH_4^+ and acidity.

● For upland forests in northern and western Britain, cloud droplet deposition represents an important input for SO_4^{2-}, NO_3^-, NH_4^+ and H^+. This is also the mechanism by which foliage is exposed to very large ion concentrations (up to 1 mM SO_4^{2-}, NO_3^-, NH_4^+ and H^+).

● Forests increase the inputs of the pollutants HNO_3, NH_3, SO_2 and ions deposited in cloud droplets relative to short vegetation. The increase varies with many environmental variables but is typically by a factor of 2 relative to short vegetation.

● Annual average SO_2 and NO_2 concentrations are similar in rural areas of the UK and increase from 1–2 ppb in northern Scotland to 10–15 ppb in parts of the Midlands and south of England.

● Inputs of sulphur and nitrogen to trees in the Midlands and the south of England are dominated by dry deposition.

● Amenity trees in cities are exposed to the largest SO_2 and NO_2 concentrations in the UK.

● Inputs of ammonia by dry deposition are the largest atmospheric nitrogen input to unfertilized vegetation over most of east and southern Britain. Inputs of dry deposited NH_3 to forests in central and eastern England, probably range from 15 to 75 kg N ha^{-1} y^{-1}.

● Concentrations of potentially phytotoxic ozone (> 60 ppb) occur as episodes throughout the UK reaching maximum hourly averages in excess of 100 ppb. The number of days experiencing >60 ppb ranges from 2–10 days per year in northern Scotland to 10–40 days in southern England.

● In episode conditions, upland, windy sites continue to be exposed to large concentrations of O_3 for 24 hours a day unlike sites on low ground. Thus such upland sites are exposed to longer duration of high O_3 concentrations than adjacent low ground.

The Pollution Climate of the United Kingdom and Tree Health

The term pollution climate has been deliberately chosen as it incorporates:

i) the major phytotoxic atmospheric pollutants eg sulphur dioxide (SO_2) oxides of nitrogen (NO_x) and ammonia (NH_3) and atmospheric reaction products of these pollutants such as sulphate (SO_4^{2-}) nitrate (NO_3^-) ammonium (NH_4^+) and ozone (O_3).

ii) Aspects of the physical climate which influence the concentration and/or dose and combination of the above pollutants to which the trees are exposed.

General Characteristics of Pollutant Deposition on Forests

The tall, aerodynamically rough morphology of trees leads to very efficient exchange of heat and mass with the atmosphere and therefore to large potential rates of deposition of pollutants into forests. The effect of forests on individual pollutants is discussed in more detail below but in a general sense the pollutants fall into two groups. In the first group are those for which rates of deposition on forests are close to the maximum value provided by atmospheric processes (Vg_{max}) and which represent uptake by the leaf and branch sur-faces. These are the gases nitric acid (HNO_3), hydrogen chloride (HCl), ammonia (NH_3) and in some circumstances SO_2 and cloud droplets. Nitric acid for example, exhibits no surface or canopy resistance (Fig. 2.1) and the rates of deposition depend only on rates of turbulent and diffusive transfer to the trees. The importance of this effect may be quantified using a maximum deposition velocity (Vg_{max}) estimated from a knowledge of the typical rates of transport of gases, heat and momentum between vegetation and the atmos-phere. The values in Table 2.1 contrast maximum deposition velocities of HNO_3 on short vegetation (height 0.5 m) and forests (height 10.0 m) and show that deposition velocities of the very reactive pollutants HNO_3, HCl, NH_3, are larger on forests by a factor of about 5.

Table 2.1: Deposition of HNO_3 on moorland and forest vegetation

Vegetation	Moorland	Forest
Height (m)	0.5	10.0
Windspeed m s^{-1} (3 m above canopy)	4.0	4.0
Maximum deposition velocity cm s^{-1}	3.0	15.0

The second category includes gases which are absorbed via stomata and are not absorbed readily by external foliar surfaces. For these gases the rates of dry deposition are similar on both forests and shorter vegetation (Fig. 2.2) and under the control of stomatal conductance exhibit marked diurnal and seasonal cycles. This group includes NO_2, O_3 and SO_2 (in the absence of gaseous NH_3).

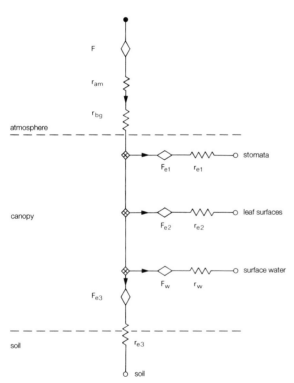

2.1 A resistance analogue for pollutant uptake by crop canopies. The total resistance of r_t is the sum of atmospheric resistances r_{am} + r_{bg} and canopy resistance r_e. Canopy resistance is determined by up to four parallel paths, stomata r_{e1}, external plant surfaces r_{e2}, surface water r_w, and soil r_{e3}; corresponding fluxes are denoted by the prefix F.

Topography

As air is lifted up hillsides in the UK, the sub-micron particles containing SO_4^{2-}, NO_3^-, NH_4^+ and H^+ are readily activated into cloud droplets close to

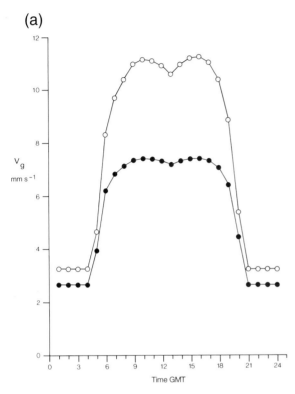

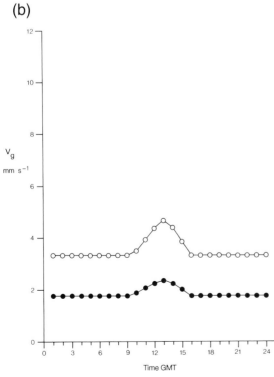

2.2 Deposition of SO₂ to coniferous forest (○) and grassland (●) in: (a) June; (b) December, 56° North.

cloud base. The droplets grow rapidly as they are lifted further and the droplets in these low level orographic clouds are commonly found in the range 3 to 20 μm radius. Such droplets are readily captured by trees whereas the aerosols at low elevation (< 300 m) from which the droplets are produced are generally too small to be captured readily by forests. Similarly, gases may be incorporated into these orographic clouds. Thus the geographical location and the altitude of forests greatly influence the form of the pollutant and the rate at which it is deposited.

Gaseous Deposition

Sulphur dioxide

The annual mean concentrations of SO₂ over rural Britain lie in the range 2 to 15 ppb (Fig. 2.3), with most areas in the range 5 to 10 ppb. These concentrations are small by comparison with the concentrations necessary to cause visible foliar lesions. However the use of mean concentrations conceals the variability from day to day or hour to hour. Primary pollutants such as SO₂ show a log-normal frequency distribution with the bulk of the observations at small concentration and a long 'tail' of infrequent events with large concentrations. For a typical rural location with a geometric mean concentration of 10 ppb and a geometric standard deviation of 2.0, approximately 70% of observations would lie between 5 ppb and 20 ppb and 90% between 2.5 and 40 ppb, so that concentrations exceeding 40 ppb would only occur for 5% of the time. These features of the concentration data are shown in Figure 2.4.

The dry deposition of SO₂ onto vegetation has been studied more extensively than any other pollutant and the major features of the process are well understood.

Stomata are a major sink for SO₂ and except in the presence of large ambient ammonia concentrations or surface water, leaf surface uptake is a minor component of the total. The map of dry deposited SO₂ in the UK (Fig. 2.5) has been

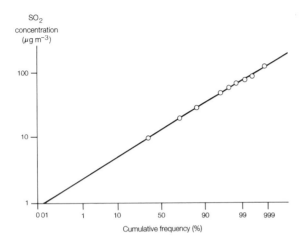

SO$_2$
concentration
(μg m^{-3})

2.4 Frequency distribution of sulphur dioxide concentration: log-probability plot. (From Fowler and Cape 1982.)

produced using a resistance model of the type described by Monteith and Unsworth (1990) and is described in more detail in the Review Group on Acid Deposition report published in 1990 and by Fowler *et al.* (1989a). The values presented are the computed dry deposition for each 20 x 20 km square of the county. The model has incorporated the differences between forests and other vegetation and has been scaled within each square for the proportions of each of the major land uses (urban, arable, permanent grass, moorland and forest).

One significant uncertainty in the SO$_2$ dry deposition estimates is the effect of gaseous ammonia. It is generally believed that in the presence of gaseous ammonia at concentrations comparable with those of SO$_2$ the two gases react on foliar surfaces (co-deposition) permitting the deposition of SO$_2$ to approach the upper limit set by turbulent transfer (Vg$_{max}$). If this is found to be generally applicable then rates of SO$_2$ deposition will be larger than those shown in Figure 2.5, and for forests in particular where the values of Vg$_{max}$ reach 15 cms^{-1} the inputs of SO$_2$ by dry deposition may be much larger in areas with similar concentrations of NH$_3$ and SO$_2$. However there is as yet insufficient experimental evidence to justify modification of the estimates in Figure 2.5.

Nitrogen dioxide and nitric oxide

The annual mean concentrations of nitrogen dioxide throughout the UK are shown in Figure 2.6. The largest rural values are found in the Midlands and south east England with values in the range 10 to 20 ppb. In many areas the concentrations of NO$_2$ exceed those of SO$_2$, but both are of a similar order. These data were obtained using diffusion tube techniques which provide average values for a two-week exposure. The short term (hourly) variability in NO$_2$ concentration has therefore to be obtained from continuous monitors. Using data from continuous instruments the NO$_2$ concentrations show an approximately log-normal distribution but with a smaller geometric standard deviation than for SO$_2$ by typically 20%. This is a consequence of NO$_2$ being largely a secondary pollutant, formed in the atmosphere by reaction of the primary pollutant nitric oxide with ambient ozone.

The deposition of NO$_2$ on vegetation is now generally believed to depend on both NO$_2$ concentration and the sink strength for nitrogen in the canopy. At concentrations of NO$_2$ less than 5 ppb, there is negligible deposition onto semi natural vegetation such as unfertilized forest canopies and moorland vegetation. At many clean air sites the small upward nitric oxide flux from soil denitrification is similar to, or greater than any NO$_2$ deposition. For fertilised cereal crops, the NO$_2$ uptake approaches the maximum permitted by stomatal conductance and at concentrations of 5 to 15 ppb which includes large arable, pasture and urban areas of the Midlands and south of England, NO$_2$ is deposited at typically Vg 1 to 4 mm s^{-1}. The data from which these estimates have been produced include a very detailed study of NO and NO$_2$ exchange over grassland, but the estimates are nevertheless, still speculative since the published literature of NO$_2$ deposition velocity remain very variable in sign as well as magnitude.

Because of the obvious problems of using micrometeorological methods to determine NO$_2$ deposition to forests, estimations are usually made by cuvette techniques. Deposition values for pine

Table 2.2: NO₂ Deposition onto Forest, Moorland and Grassland

Site	Vegetation	Concentration ppb	Deposition Velocity of NO₂ (mm s⁻¹)	Reference
Sweden Jadraas	Scots pine forest	> 10.0	4.0–7.0	Johansson, 1987
Sweden Jadraas	Scots pine forest	1.0–3.0	1.0–2.0	Johansson, 1987
UK	Acidic moorland	0.5	1.0–2.0	Fowler *et al.*, 1989a
UK East Anglia	Marshland pasture	10.0	4.5	Hargreaves *et al.*, 1991

forests at high and low NO₂ concentrations estimated from recent cuvette studies in Sweden (Johansson, 1987) are given in Table 2.2 together with micrometeorological measurements of NO₂ deposition to moorland and marshland pasture in the UK. At the concentrations of NO₂ found in large areas of coniferous forest in Sweden and some areas in the UK, NO₂ uptake is likely to be very small.

The nitric oxide concentrations in urban areas represents a significant fraction of ambient NOₓ but outside the cities and away from major roads they are a minor fraction. The deposition of NO on to plants is very slow with deposition velocities in the range 0.1 to 0.5 mm s⁻¹. It therefore represents a small fraction of deposited NOₓ on to plants at most sites.

Ammonia

The ambient concentrations of ammonia in the UK have been measured using diffusion tube methods (Hargreaves and Atkins, 1988). These measurements have been used to provide a map of average ambient concentrations (Fig. 2.7). The concentrations are largest in the border counties of England and Wales and in parts of East Anglia

where levels exceed 10 ppb. In much of the north and west of Britain concentrations are generally in the range 0.1 to 3 ppb. The frequency distribution for NH₃ would be expected to be log-normal as this is a primary pollutant from largely ground level sources which is diluted by turbulence. The measurement technique however supplies insufficient data to provide reliable estimates of the short term variability.

Ammonia is readily dry deposited on vegetation. For moorland vegetation, forests and unfertilized grassland, rates of dry deposition of NH₃ approach the maximum values provided by turbulent transfer. The inputs of nitrogen to these land surfaces via dry deposited ammonia may therefore be large and may greatly exceed those from deposition of NO₂ and HNO₃. Over agricultural vegetation the NH₃ fluxes are more complex since both emission and deposition fluxes have been observed. As atmospheric inputs of nitrogen from ammonia deposition are large on forests, the proximity of woodland and hedgerows to major agricultural sources of ammonia is a very important issue.

Physiological studies have shown (Farquhar *et al.*, 1980) the existence of a substomatal 'compensation point' concentration of NH₃ in plants, which may be of the same order as atmospheric

8

concentrations. Ammonia deposition occurs only when atmospheric concentrations exceed the 'compensation point'. When they are lower than the compensation point, NH_3 emission occurs. Vegetation fertilised with large quantities of inorganic fertiliser also appears to exhibit a 'compensation point' and a bi-directional exchange of NH_3 has been observed (Sutton, 1990), with warm dry conditions favouring emission and cool wet conditions favouring deposition. There is no field evidence over unfertilized vegetation that such compensation points are significantly greater than zero, while for cropland, insufficient field measurements have been completed to show the seasonal or diurnal variation in compensation point or to its dependence on the nutritional status of the plants. It is possible therefore to estimate the deposition fluxes over moorland and forests from the assumption of zero surface resistance. An example of such an exercise is provided in the last section of this chapter.

Ozone

The ambient concentrations of ozone exceed those of the two other major phytotoxic gases SO_2 and NO_2 and are typically in the range 25 to 30 ppb. Variability in mean ozone concentrations between sites is largely the consequence of processes which deplete ozone in the surface layers, for example, dry deposition beneath the nocturnal temperature inversion or chemical destruction of ozone by nitric oxide. There is little evidence that these typical background concentrations represent a threat to tree health. The primary interest in O_3 as a phytotoxic pollutant is its presence at concentrations in excess of 60 ppb during photochemical episode conditions. Such concentrations have been shown in controlled studies to be damaging to a number of physiological processes in trees ranging from water relations and photosynthesis to frost hardening. The frequency of photochemical episodes varies considerably between years in Britain and is closely linked with the meteorological conditions. Photochemical O_3 episodes occur under warm (ambient temperatures

in excess of 20°C) sunny conditions with sources of the photochemical oxidant precursors upwind, which for the UK generally, means winds from the south and east. Such conditions have produced between 50 and 200 hours above 60 ppb at most sites in southern England during the last two warm years. Further north the frequency of episodes declines and the maximum values achieved are smaller (Fig. 2.8).

The altitude of the land also has a marked effect on the length of exposure to episode levels of ozone since low altitude sites generally lose the episode concentrations during the night due to deposition at the ground in low windspeed conditions. At high elevation as for example at Great Dun Fell during the episode 24–30 April 1987 (Fig. 2.9) the higher wind speeds generally prevent the development of a stable layer close to the ground and episode concentrations of O_3 remain for a longer period. As a result of these effects, the high elevation forests (250 m to 600 m) in the UK receive a longer exposure to episode concentrations of O_3 when they occur. It is important to note however that there is also a

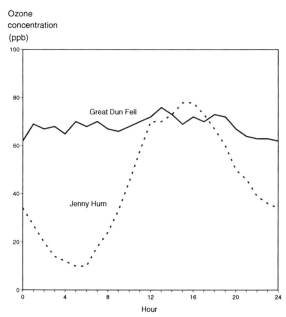

2.9 The diurnal variation of ozone concentration at Great Dun Fell (847 m a.s.l.) and Jenny Hurn (4 m a.s.l.) averaged over the period 24–30 April, 1987.

gradient of decreasing frequency of episodes from south east England to north west Scotland.

Because of the combination of geographic location and altitude, areas of the uplands may experience the longest annual exposure (number of hours) to concentrations of O_3 in the range 40–60 ppb even though forests in the south of England are exposed to the largest concentrations of O_3.

Nitric acid vapour

The oxidation of NO_2 in the atmosphere to HNO_3 provides a background concentration of this very reactive and readily deposited gas. Concentrations in the atmosphere over the UK are generally in the range 0.1 to 1.0 ppb and do not constitute a direct threat to tree health. However, the very efficient exchange of gases over forests leads to very large deposition rates of HNO_3 onto the trees, thus making the contribution of HNO_3 to deposited nitrogen on forests a significant component.

Urban areas

The concentrations of the primary pollutants SO_2, NO, and the secondary pollutant NO_2 are generally largest in urban areas. Concentrations of NO in streets and major cities frequently exceed 100 ppb while those of NO_2 commonly reach peak concentrations in the range 30 to 50 ppb. Concentrations of SO_2 in urban areas have declined steadily during the last three decades and are currently in the range 10 to 30 ppb for major cities. Urban trees are therefore subject to the largest exposure to ambient levels of SO_2 and NO_2. However the urban/rural contrast has changed in recent years with urban areas generally becoming cleaner, especially with respect to SO_2 and smoke while rural areas have progressively become more polluted by NO_x, especially by motor vehicle emissions. Table 2.3 shows concentrations of NO_2, NO and SO_2 in 5 major UK cities.

Table 2.3: Urban concentrations of NO_2 and SO_2 in the U.K.

Sites	SO_2 ppb	NO_2 ppb
Background sites	12	16
Intermediate sites	13	19
Near Road sites	13	22

NO_2 data are (6 months, July–December) means of diffusion tube monitoring data from 363 urban sites (background sites are more than 50 m from a busy road, intermediate sites are less than 50 m from a major road and 'near road' include both kerbside sites and locations a few metres from busy roads). SO_2 data are 6 month mean values from urban smoke and SO_2 monitoring networks. The data are taken from Bower *et al.* (1989).

Wet deposition

The deposition of major ions in rainfall in the UK has been extensively monitored during the last 4 years and the results are described in detail in the report of the UK review group on acid deposition (RGAR, 1990).

The major impacts of wet deposition on forests occur through the soil mediated indirect effects of wet deposited SO_4^{2-}, NO_3^- and acidity and the physiological effects of intercepted cloud water which in polluted air may contain concentrations of SO_4^{2-} and NO_3^- in excess of 1000 eq l^{-1}.

The wet deposition of major ions is largest in the high rainfall areas of the west and north of Britain. The earlier published values for wet deposition in the high rainfall areas of the north and west did not satisfactorily incorporate the effects of hills on rainfall composition and were therefore underestimated. The revised maps presented here for SO_4^{2-}, NO_3^- and H^+ (Figs. 2.10, 2.11 and 2.12) include the effects of hills on rainfall amount and composition. The inputs of sulphur by wet deposition to the ground dominates the input budget over the country as a whole and especially in the areas of high rainfall. The combination of high rainfall and large concentrations of the major ions in rain in areas of the Pennine Hills and Cumbria are a notable feature. The majority of the forest areas in the UK are located in the regions in which wet deposition represents the largest input mechanism for SO_4^{2-}, NO_3^- and acidity.

Cloud droplet deposition

The capture of cloud droplets from hill cloud by forests is a particularly efficient process and rates of deposition by this mechanism which is analogous to the dry deposition of reactive gases may reach deposition velocities of 10 to 15 cm s^{-1} ($\approx Vg_{max}$). The ionic composition of cloud water on hills generally shows concentrations to be larger than rain by a factor of 2 to 8 (Table 2.4). No precise enhancement in concentration is possible because at any given site the large change in liquid water content of cloud with altitude on hills (Figs. 2.13),

causes large changes in concentration, especially close to cloud base. Although the inputs of pollutants as a fraction of the total input is small (<10%) at most sites, it becomes important at high elevation (> 400 m). Thus, a combination of efficient capture of the polluted cloud water by forests and large concentrations of pollutants in the droplets make this deposition pathway an important aspect of the pollution climate for studies of tree health. The concentration of SO_4^{2-}, NO_3^-, NH_4^+ and H^+ observed in cloud water in upland UK (and at high altitude sites in Europe and eastern North America) have been shown in controlled conditions to cause a range of physiological effects on Red spruce and Norway spruce.

Figure 2.14 shows for Britain the magnitude of cloud water deposition of sulphur as sulphate. Only in the high land in the north and west do the inputs represent a significant component of the total. The range of concentrations of major ions in cloud water are known only for 2 sites in the UK, Dunslair Heights in southern Scotland and Great Dun Fell in Cumbria, so that for most areas the concentrations of major ions to which trees on hills are exposed is uncertain.

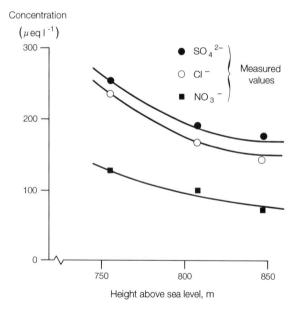

2.13 Variation in concentration of major anions in cloud, as a function of altitude, compared with that predicted from a calculation of adiabatic water content. (Taken from RGAR 1990). (_____ Predicted change with altitude. For each ion, the measured values actually fall on or close to the calculated curves.)

Table 2.4:   Ratio of concentrations in cloud/rain (μeq l^{-1})

H^+	NH_4^+	Cl^-	NO_3^-	SO_4^{2-}
3.9	2.4	2.6	2.8	2.0

Mean of 11 precipitation and cloud events at summit of Great Dun Fell (847 m) during spring 1985.

Pollution deposition and forests

The subject of this chapter is the pollution climate of the UK as it relates to aspects of tree health discussed in later chapters. Trees however, by virtue of their aerodynamic properties also greatly influence the input of pollutants to the ground. For example, in considering the afforestation of moorland sites in the border counties of England and Scotland, modelling studies have shown that planting a forest at Kielder has increased the atmospheric sulphur and nitrogen inputs by a factor of approximately 1.5 and 2.0 respectively (Figure 2.15). The effect is largely the result of the rapid deposition of nitric acid vapour, ammonia and cloud water onto forests. The greater ambient ammonia and nitric acid concentration in southern England would lead to even larger inputs of atmospheric nitrogen in these areas.

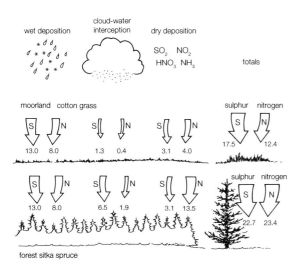

2.15 Estimated atmospheric inputs of sulphur and nitrogen (kg ha^{-1} y^{-1}) at Kielder Forest (6 x 10^4 ha, 300 m a.s.l., 1500 mm rain). (From Fowler *et al.*, 1989a.)

Summary

In describing the major features of the pollution climate of the UK and its relevance to tree health, the dominant effects of climate and the location of the major areas of commercial forest are a notable feature.

1. For the large areas of Sitka spruce plantations in northern England and southern and western Scotland, inputs of sulphur and nitrogen are dominated by wet deposition. At elevations above 300 m, cloud water deposition is also an important component of the input budget and results in the exposure of foliar surfaces to the largest concentrations of SO$_4^{2-}$, NO$_3^-$, H$^+$ and NH$_4^+$. In most of these areas ambient concentrations of SO$_2$, NO$_2$ and HNO$_3$ are all very small (<5 ppb). Ozone episodes are relatively infrequent, and typically occur on less than 10 days each year, but at high elevation (>300 m) the duration of exposure to the episode ozone concentrations is considerably enhanced (by typically a factor of 3) relative to low ground.

2. For forest and amenity trees in rural areas of the Midlands and the south of England, inputs of sulphur and nitrogen are dominated by dry deposition and especially in the case of ammonia these may be very large (up to between 50 and 100 kg N ha^{-1} y^{-1}). Concentrations of SO$_2$ and NO$_2$ lie in the range 10 to 20 ppb for each in the most polluted areas and for typically 5% of the time concentrations in excess of 40 ppb are expected. Exposure to episode concentrations (>60 ppb) of ozone is larger than in areas of the north and lie in the range 50 to 200 hours per year.

3. For amenity trees in urban areas, inputs of sulphur and nitrogen are dominated by dry deposition. Concentrations of SO$_2$ and NO$_2$ in the largest cities are typically in the range 10 to 30 ppb and exposure of trees to NO$_2$ and SO$_2$ concentrations of 50 ppb at some sites is common, but urban areas also show very large spatial variability in these concentrations. Exposure to ozone episode concentrations in cities is smaller than in the surrounding countryside as a consequence of ozone depletion by nitric oxide.

12

3

Surveys of Tree Health

Conclusions

- Consistent surveys of the crown condition of trees in the UK have only been carried out for the past four years. Major progress has been made in assessment methods.

- It is not possible to compare the crown condition of trees in the UK and Europe with confidence in the earlier surveys.

- The four years of surveys are insufficient to detect long-term trends in crown condition.

- There is no indication of a widespread decline in the crown condition of trees in the UK up to 1990, but beech at some sites in southern Britain show evidence of recent deterioration in condition.

- Between 1990 and 1991, the condition of broadleaved trees surveyed in the UN-ECE surveys at UK sites showed a marked deterioration.

- The most recent UN-ECE survey of tree health in 1991 shows the condition of UK trees to be generally poorer than in other European countries.

- When compared to an ideal tree, in 1990, 58% of Sitka spruce in the UK had a crown density loss of more than 20% and 10% had a loss of more than 50%. The corresponding figures for Norway spruce were 50% and 4%, for Scots pine were 62% and 6%, for oak were 66% and 4% and for beech were 62% and 5%.

- The surveys have indicated that a number of factors, such as stand density and age significantly affect crown condition.

- Surveys are not intended to conclusively demonstrate a causal link between the crown condition of trees in the UK and air pollution, but they are important for establishing whether a problem is present.

Tree Surveys

In the late 1970s and early 1980s, various reports of forest decline and forest death emerged from central Europe. Two species appeared to be particularly affected; namely silver fir and Norway spruce. The reports showed that the health of these two species was deteriorating in a number of areas and several predictions of complete forest loss were made. The damage was attributed to the effects of air pollution, with "acid rain" being singled out as the cause. The problem resulted in the establishment of surveys of forest health throughout Europe, the primary aim of which was to determine the extent of any decline in health. This aim was quickly extended to include the monitoring of forest health on an annual basis.

Formal surveys of tree health were started in Great Britain in 1984 by the Forestry Commission. From the start, the British survey differed from European surveys in that an attempt was made to identify the possible role of air pollution in determining tree health. The 1984 investigation examined three tree species: Sitka spruce (*Picea sitchensis* (Bong.) Carr.), Norway spruce (*Picea abies* (L.) Karst.) and Scots pine (*Pinus sylvestris* (L.). Sampling was based on a factorial design, with approximately 100 plots located in plantations throughout Britain (Binns *et al.*, 1985). The sampling design was intended to provide a contrast between areas with high and low rainfall, high and low sulphur deposition and high and low altitude. The study was repeated in 1985, with minor modifications (Binns *et al.*, 1986). In both years,

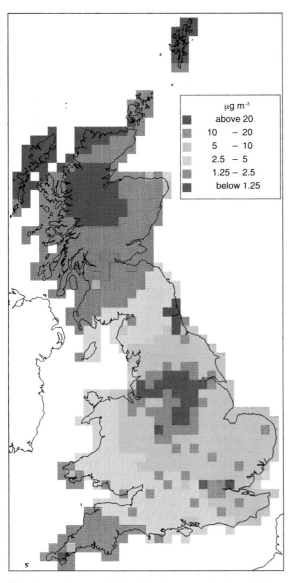

	μg m⁻³
	above 20
	10 — 20
	5 — 10
	2.5 — 5
	1.25 — 2.5
	below 1.25

2.3 Annual mean sulphur dioxide concentration (ppb), 1988.

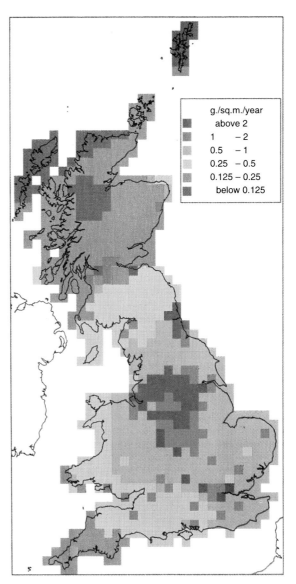

	g./sq.m./year
	above 2
	1 — 2
	0.5 — 1
	0.25 — 0.5
	0.125 — 0.25
	below 0.125

2.5 Annual dry deposition of sulphur dioxide (g S m⁻²), 1988.

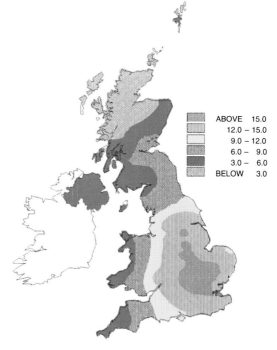

ABOVE	15.0
12.0 —	15.0
9.0 —	12.0
6.0 —	9.0
3.0 —	6.0
BELOW	3.0

2.6 Annual mean nitrogen dioxide concentration June 1987 to May 1988 (ppb).

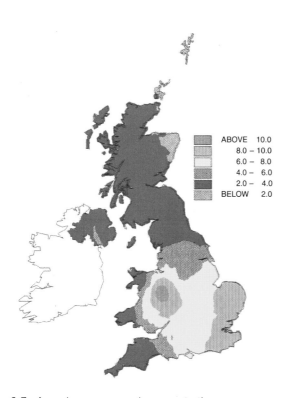

2.7 Annual mean ammonia concentration
June 1988 to May 1989 (ppb).

2.8 Number of hours during which ozone
concentration exceeded 60 ppb during April–September
1989.

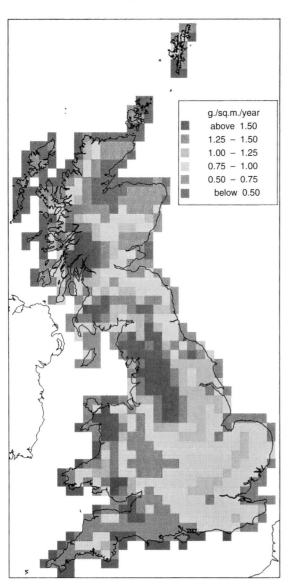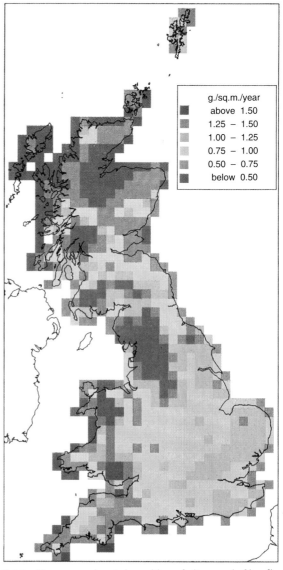

2.10 Mean annual wet deposition of sulphur (g S m⁻²)
1986–1988.

2.11 Mean annual wet deposition of nitrogen (g N m⁻²)
1986–1988.

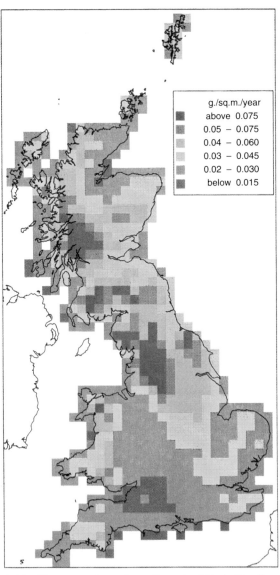

2.12   Mean annual wet deposition of acidity (g H m⁻²) 1986–1988.

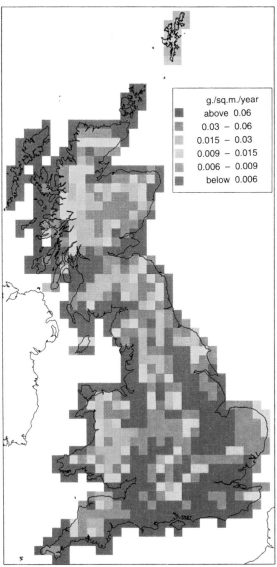

2.14   Annual mean cloud water deposition of SO₄-S (g S m⁻²) 1986–1988.

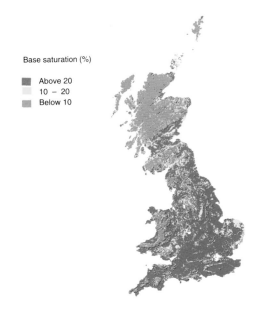

5.8   Soils of Great Britain classified on the basis of base saturation.

6.2 Extent of foliage damage in eastern England in September/October 1989.

6.3 Percentage trees affected with ash dieback; sample plot data interpolated on a 20 km x 20 km grid. (From Hull and Gibbs, 1991).

>50%
40% – 50%
30% – 40%
20% – 30%
10% – 20%
<10%

two indices believed to reflect tree health were assessed: crown density and crown discoloration. On the basis of these indices, neither survey identified a widespread problem although it was clear that specific problems existed at a number of sites. This conclusion contrasted markedly with the results of a survey by the Friends of the Earth (Rose and Neville, 1985) which concluded that typical symptoms of acid rain damage to trees were widespread in Britain. The difference in interpretation can be attributed to the way in which crown density and discoloration were interpreted. While both may be caused by acid rain, a variety of other factors can also produce the same symptoms, making it impossible to pinpoint acid rain as the sole cause.

In 1986, standardization of the observations made by different surveyors in the Forestry Commission survey was improved (Innes *et al.*, 1986). This was followed in 1987 by the establishment of a training course, based on that used in southern Germany (Schöpfer, 1985). At the same time, the standards used to assess crown condition were revised to bring the British assessments into line with those adopted in Switzerland (Bosshard, 1986). The changes in survey standards had a marked effect on the results. Comparability studies undertaken in 1986 and 1987 indicated that surveyors were underestimating reductions in crown density relative to their German counterparts and also that there were marked differences between the surveyors. The re-calibration to Swiss standards resulted in a recorded deterioration in the condition of trees between 1985 and 1987, although there is no evidence that this was anything other than an artefact of the re-calibration. The re-calibration, however, introduced an uncertainty in comparisons of data obtained before and after 1987. To provide a safe interpretation, the analysis was therefore confined to data obtained from 1987 onwards.

Another change made to the Forestry Commission surveys in 1987 was the inclusion of two broad-leaved species, oak (*Quercus petraea* (Mattuschka) Lieblein, *Q. robur* L. and hybrids between the two) and beech (*Fagus sylvatica* L.). Plot numbers were initially small but have been gradually expanded and, in 1991, the number of plots was approximately equal for the five main species investigated by the Forestry Commission.

Considerable progress has been made in the assessment of crown condition since the early surveys. In 1984, only crown density and crown discoloration were recorded. By 1989, over thirty different measures of crown condition were being recorded for each species, enabling a much more detailed assessment of crown condition than is possible in the standard European surveys. Work has been undertaken on the problems of calibrating different teams of observers (Innes, 1988a), on the analysis and presentation of annual data (Innes and Boswell, 1990), on the spatial analysis of forest health data (Innes and Boswell, 1989a) and on the use of multiple indices of crown condition (Innes, 1990a). The effect of this work is that an accurate picture of crown condition can now be established so that very small changes in condition can be identified and quantified as they occur. For example, small changes in the percentages of trees showing signs of poor condition can now be confidently assessed.

A number of other investigations of tree health have been undertaken. These have mostly concentrated on two species, beech and yew (*Taxus baccata* L.), although Scots pine and oak have also been included in some. In most cases, these surveys have examined trees in different situations to the Forestry Commission surveys. For example, the survey of beech conducted by Imperial College concentrated on sites of conservation interest. Because of the number of different surveys, the results for each species have been described individually below.

In the majority of surveys, assessments have concentrated on crown density, crown discoloration and crown architecture. The relationship between these and injury by pollution is unclear. Distinctive symptoms of acute injury by pollutants are known to occur in Britain. Although these are mainly restricted to areas around point sources, exceptions occur *(see Chapter 6)*. The surveys of tree health have therefore concentrated on identifying symptoms that might reflect chronic injury. As the surveys are undertaken from the ground, certain symptoms, such as needle flecking caused by ozone injury, may be missed (Muir and Armentano, 1988). In addition, surveys have concentrated on visible symptoms whereas there is a considerable amount of evidence that the first response to stress may involve changes in the growth rate of individual trees or the yield of stands, in the absence of visible injury. Annual records of growth rates were started in the Forestry Commission surveys in 1989.

Sitka spruce

Sitka spruce is a widely planted species which was introduced from the west coast of North America in 1831 (Lines, 1987). It favours maritime conditions and, in Britain, grows extremely well in the north and west of the country. In this area, its growth appears to be determined locally by its growing environment, particularly the soil and climatic conditions (Worrell, 1987). Because of its fast growth, it is widely planted and it is now the most common tree species in Britain, constituting 28% of the forested area in 1980 (Locke, 1987). Its importance as a commercial tree species has meant that there has been a great deal of research on the species (e.g. Henderson and Faulkner, 1987).

The Forestry Commission is the only body that has undertaken surveys of the condition of Sitka spruce across Britain. As the results of the 1984–1986 surveys are not comparable with those obtained subsequently, only the results for 1987–1990 have been presented here (Table 3.1). The data for the crown densities of the trees suggest that the majority of trees have crown density reductions (defined as the departure from a fully-foliaged tree) of between 10% and 49%. On the basis of only four years' data, it is impossible to identify any temporal trends, but there are clear variations from year to year. These have been largely attributed to the influence of the green spruce aphid (*Elatobium abietinum* Walker;

Hemiptera) (Innes and Boswell, 1990). Infestations by this insect have had a severe effect on many trees and in 1989, 4% of the trees assessed in the survey had only one or two years' needles present instead of the normal 8 to 12 years'.

The thin crowns of some of the trees can be related to the age of the trees and their growth rates. In both Sitka spruce and Norway spruce, crown density decreased with increasing age, a trend that is also apparent for older broadleaves. Faster-growing trees have more widely spaced branches, resulting in more transparent crowns. This has been examined in detail (Innes and Boswell, 1990a) and although rate of growth probably explains a substantial proportion of the crown thinness at density reductions of 40% or less, trees with higher levels of thinness cannot be readily explained in this fashion. No relationship has been identified between crown density and tree dominance for any of the species, contrasting markedly with results from Germany (Krause *et al.*, 1983; Prinz and Krause, 1989), although this may reflect differences in the way dominance is classified in Germany and Britain.

The pattern of defoliation provides an important diagnostic tool when examining crown thinness. The proportions of trees in each of a number of types in 1990 are presented in Table 3.2. The most frequent type of defoliation involved the uniform loss of needles throughout the crown. In Sitka spruce, needle loss at the base of the crown, the second most common type of defoliation, was usually associated with infestations by the green spruce aphid.

Other indices of crown condition have been recorded. These include assessments of tree and branch form, patterns of shoot death and evidence of mechanical and other forms of damage to the crown, stem and butt. Full details are provided in Innes and Boswell (1989a and 1990a). Discoloration of needles has been relatively rare. Where it occurs, it is mainly restricted to older needles.

No relationship has been found between the crown condition of Sitka spruce and the patterns of climatic or pollutant variables experienced in Britain. This contrasts markedly with work which

Table 3.1: Comparison of tree crown density results for the period 1987–1989. Class 0 represents 0–10% reduction in crown density, class 1 represents 11–20% reduction, class 2 represents 21–30% reduction and so on. Data are given only for trees assessed in all four years. (n = total number of sampled trees).

Species	Year	Crown density class									
		0	1	2	3	4	5	6	7	8	9
Sitka spruce	1987	15	23	24	19	11	5	2	0	0	0
n = 1064	1988	8	24	26	22	14	5	1	0	0	0
	1989	11	23	24	20	10	5	3	2	2	0
	1990	17	25	23	18	7	4	2	2	1	0
Norway spruce	1987	21	24	24	15	9	4	2	1	0	0
n = 1431	1988	20	26	24	17	7	4	1	0	0	0
	1989	25	28	22	14	6	2	1	1	0	0
	1990	21	29	25	15	6	2	1	0	0	0
Scots pine	1987	18	23	25	19	8	3	1	1	1	1
n = 1348	1988	7	22	32	21	10	4	1	1	1	1
	1989	11	30	28	17	8	3	2	1	1	1
	1990	10	28	34	17	5	2	1	1	0	1
Oak	1987	8	15	22	30	14	6	4	1	0	0
n = 724	1988	4	16	30	29	13	6	2	1	1	0
	1989	6	21	30	27	10	3	1	0	0	0
	1990	10	24	32	22	8	3	1	0	0	0
Beech	1987	8	23	26	27	12	3	1	0	0	0
n = 642	1988	10	24	32	23	9	2	0	0	0	0
	1989	19	30	29	16	4	1	1	0	0	0
	1990	17	21	26	20	11	3	1	0	0	0

has revealed a close relationship between the growth of Sitka spruce and climatic variables in northern Britain (Worrell, 1987). The condition of stands is very variable, with adjacent stands showing differences in crown condition as large as stands separated by long distances (Innes and Boswell, 1989a). This suggests that the effects of local site factors override any more widespread effects that might be brought about by factors such as air pollution. However, despite screening a variety of climatic, topographic and edaphic factors, the principal environmental influences on Sitka spruce condition, as revealed by the survey, have not been identified, suggesting that genetic factors may be important.

Norway spruce

Norway spruce is the species in central Europe that has attracted the greatest amount of concern in relation to forest decline. The majority of work in Germany has focused on this species as it is much more important than silver fir (*Albies alba* Mill.), accounting for 39% of the German forest area in comparison to the 2% covered by silver fir (Anon, 1989). In Great Britain, it is much less important, accounting for only 6% of the forest area (Locke, 1987). Norway spruce is an introduced species in Great Britain, with the majority of trees planted prior to 1939 coming from Austria and Germany (Lines, 1987). Since then, eastern Europe has been the favoured source, although the rate of planting of this species has been declining in recent years due its replacement by Sitka spruce. The continental origins of the trees indicate its ecological preferences, and it appears to grow best in the drier, eastern areas of Britain.

The results of the assessments of crown density are given in Table 3.1. The species is less susceptible to the green spruce aphid than Sitka spruce, and the severe outbreaks of this pest in 1988–1989 appear to have had relatively little impact on the condition of the trees. As with Sitka spruce, the rate of growth has an important impact on the density of the crowns, particularly at the denser levels. There is no evidence of a link between the dominance of trees and their crown density, but a higher proportion of thin-crowned trees were noted in more open stands, a trend also apparent for Sitka spruce. This is similar to findings in Germany (Denstorf et al., 1984; Levin, 1985), Switzerland (Keller and Imhof, 1987) and Austria (Neumann, 1989).

The most frequent pattern of defoliation involved the uniform loss of needles throughout the crown (Table 3.2). This contrasts markedly with the most severe type of decline recorded in central Europe, which is typically associated with the development of gaps in the upper crown, sometime referred to as sub-top dying (Schröter and Aldinger, 1985; Westman and Lesinski, 1986; Gruber, 1988).

Analyses of the condition of the trees in relation to patterns of air pollution have not revealed any clear association between the two. In 1987, no relationship was identified between any of the pollutants investigated and either crown density or discoloration. In 1988, denser crowns were identified in areas characterised by a warmer and drier climate and also by higher levels of gaseous pollutants. None of the other indices of crown condition appeared to show a relationship (Innes and Boswell, 1989a).

Table 3.2: Defoliation types in spruce recorded by the Forestry Commission in 1990. Types: 0: no obvious defoliation, 1: small window in upper crown; 2: large window in crown; 3: top-dying; 4: uniform loss of needles/branches; 5: peripheral loss; 6: bottom upwards; 7: other.

	Defoliation type							
	0	1	2	3	4	5	6	7
Sitka spruce	25	1	1	0	51	0	18	4
Norway spruce	39	1	3	1	39	1	14	2

Scots pine

Scots pine occurs widely in Great Britain, being the second most common species. In 1980, it accounted for approximately 13% of the forest area of Great Britain (Locke, 1987). Although it is one of the few coniferous species that occurs naturally in Great Britain many current plantations are derived from foreign stock. Although the majority of seed used after 1919 was of British origin, the sources of the seed trees are mostly unknown (Lines, 1987). Authentic areas of natural Scots pine are largely restricted to the Highlands of Scotland where several distinct populations are known to occur (Forrest, 1980, 1982; Kinloch et al., 1986). As a result of the confusion over the origins of specific stands, it is difficult to make any generalisations about the ecological preferences of the species in Britain. Trees growing in the south and east of the country appear to favour drier, sandy sites and grow badly when planted in wet, high-altitude sites, despite the presence of native pinewoods in such areas.

There have been two studies of the condition of Scots pine in Great Britain. The largest involves the annual monitoring programme undertaken by the Forestry Commission, but the species has also been investigated by Imperial College (Power *et al.*, 1989). The two studies involve rather different populations, with the Forestry Commission concentrating on plantations and the Imperial College study focusing on isolated mature individuals growing on lowland heath in the south of England.

The Forestry Commission's results for crown density during the period 1987–1990 are given in Table 3.1. Examples of crown density are shown in Plates 3.1 and 3.2. No trend is apparent, with marked variations being observed from year to year. A small proportion of trees have very thin crowns and tree mortality has been noted during the period of the assessments. Crown density appears to be particularly poor in the south of Scotland. This has been attributed to the combined effects of the needle-cast fungus (*Lophodermium seditiosum* Minter *et al.*), scleroderris canker (*Brunchorstia pinea* (P. Karsten) Höhnel (the pycnidial state of *Ascocalyx abietina* (Lagerberg) Schlaepfer-Bernard)) and the pine-shoot beetle (*Tomicus piniperda* L; Coleoptera), although why these should affect stands in this area so severely is unknown. As with the Sitka and Norway spruce, some of the crown transparency can be attributed to the rate of growth of the trees, with the fastest growing trees generally having thinner crowns than slowly growing ones.

The densest crowns occurred in warm and dry areas which, as pointed out above, are also those areas in which the concentrations of many gaseous pollutants are highest.

The Imperial College survey showed that 66% of the trees surveyed had a reduction in crown density of greater than 25%. However, no significant relationships were found between crown desity and site-factors, climate or pollution levels. Needle loss was greater on poorly-drained soils, whilst needle browning was greater in areas with high NH_3 concentrations.

Oak

Oak is the most common of the native hardwoods currently growing in Great Britain, accounting for 9% of the forest area in 1980 (Locke, 1987). Two main species are present, pedunculate or English oak and sessile oak, although these are known to hybridize freely. The two species have rather different ecological preferences with sessile oak tolerating poorer, drier and more acidic soils (Lines, 1987). Most stands over one hundred years old are believed to be of British origin although there has been a recent tendency to import seed in years of poor acorn production (Lines, 1987).

Oak has been examined in the Forestry Commission's monitoring programme and in a limited study conducted by Greenpeace (Tickle, 1988). The latter concentrated on sites in the south of England whereas the Forestry Commission survey covers woodland oaks throughout Great Britain.

The results of the Forestry Commission surveys over the past four years are presented in Table 3.1. The sample size reported is rather misleading as the number of trees assessed each year has steadily increased and, in 1990, involved about 1700 trees. However, for comparative purposes, it is necessary to restrict the sample to those trees assessed in all four years. The data for 1990 are presented in Table 3.3 and examples of crown density loss are shown in Plates 3.3 and 3.4.

In 1990, 50% of the trees assessed in the main Forestry Commission programme had some form of dieback (Innes and Boswell, 1990a). In the majority of cases, dieback involved relatively small branches, typical of a transient response to stress. However, 7% of trees had dieback involving relatively large branches or the main stem, suggesting long-term problems.

Higher crown densities have been identified in areas with higher levels of sulphur pollution. However, as with the coniferous species, these areas are characterised by favourable climatic conditions. No correlations have been identified between pollution levels and the extent of dieback or discoloration.

18

Table 3.3: The percentage of oak and beech in each 10% crown density category in 1990. The sample sizes are substantially larger than in those used in Table 3.1 because of the extension of the programme for these two species over the last two years. The classes are the same as in Table 3.1.

	Crown density class									
	0	1	2	3	4	5	6	7	8	9
Oak n = 1740	13	21	25	25	9	4	2	1	0	0
Beech n = 864	17	21	26	20	11	3	1	0	0	0

The Greenpeace survey (Tickle, 1988) suggested that 52% of the trees surveyed had losses of crown density of 25% or more. These trees included both woodland and non-woodland specimens, making interpretation and comparison with other surveys difficult. The full Forestry Commission 1988/1989 surveys of oak found a very similar number of trees (48%) with losses of crown density above 25%. These results are of interest as the Greenpeace survey was concentrated in the south of England whereas the Forestry Commission survey covered the whole of the country. As the Commission survey indicated that the condition of trees was poorest towards the north of the country, the implication is that the Greenpeace survey revealed trees in rather worse condition than the Forestry Commission survey. The inclusion of non-woodland trees, which are generally believed to be subject to more environmental stresses than woodland trees, is a possible reason for the difference.

Beech

Beech is relatively unimportant as a commercial forest species in Britain, covering only 4% of the forested area (Locke, 1987), but it has been singled out for attention because of widespread reports of declines in its health in Europe and because of its conservation and amenity import-

ance. It is widely believed to be a native species, although its natural range probably only extends over south and east England and southeast Wales (Godwin, 1975; Lines, 1987). Many planted stands are of foreign origin as the availability of mast is very variable from year to year (Edlin, 1962). As a species, it is tolerant of a wide range of soil and climatic conditions and has been widely planted on thin calcareous soils where these occur in southern England (Lines, 1987). Surveys of beech condition have been conducted by the Forestry Commission (as part of their main survey and as a special survey (Lonsdale, 1986a, 1986b)), Imperial College (Power *et al.*, 1989; Power and Ashmore, 1991), Greenpeace (Tickle, 1988) and Friends of the Earth (Rose and Neville, 1985).

The main monitoring programme of the Forestry Commission has relatively few beech sites as the species has been covered by other studies. The results for trees assessed in each of the four years during the period 1987 to 1990 are given in Table 3.1 and the results for 1990 are given in Table 3.3. One of the most obvious trends is that the proportion of trees with thin crowns increases with increasing openness of the stand, a phenomenon that has also been observed in central Europe (Keller and Imhof, 1987). This has important implications given the opening up of many semi-natural stands that occurred following the 1975/1976 drought and the 1987 storm.

A variety of different measures of crown form in beech exist and several of these have been investigated. The Forestry Commission experienced difficulties with the system developed by Roloff (1985a, 1985b) and a more detailed classification system, described by Westman (1989), has been adopted. The Roloff system is based on the changes in crown architecture that occur as a result of changes in the relative growth rates of primary and secondary shoots. The Westman system is primarily concerned with the development of gaps within the crown canopy. It was initially developed for birch but has since been extended to include other broadleaves. The results obtained for trees assessed in 1990 are presented in Table 3.4. Interpretation of these figures is difficult as there are no previous records with which they can be compared. However, they suggest that a high proportion of trees have small to large gaps in the crown whereas relatively few have major gaps.

Table 3.4: Percentages of beech in specific crown pattern categories. 0: 0-15% loss of density; 1: no clear pattern; 2: whole or part of crown transparent due to small gaps in the foliage; 3: gaps in the lateral branch system; 4: main branches bare almost to or including tips; 5: dominantly large gaps with leaves grouped at branch tips and possibly small groups on the branches; 6: whole or part of crown without leaves; 7: other.

Type							
0	1	2	3	4	5	6	7
25	10	35	20	5	0	1	4

Crown dieback was present on 44% of trees. In 41% of these, dieback was restricted to relatively small branches, suggesting that the trees have responded to some short-term factor.

Average crown density decreases noticeably towards the north and west of the country, a trend that is entirely consistent with the natural range of beech. However, the trend also occurs along a known pollution gradient, so that the densest trees occur in areas with the highest levels of gaseous pollutants and the warmest climates.

The other study by the Forestry Commission, reported by Lonsdale (1986a, 1986b), was conducted for two consecutive years to find whether there was any sign of widespread poor health as reported from Germany. Its scoring system allowed specific recognition of biotic and site-related damage, and when these factors were taken into account there was very little incidence of crown density or colour that could be judged unusual. Geographic effects could not be analysed since it was not possible to avoid confounding them with possible effects of tree age. Subsequently, a more detailed study of shoot extension was undertaken at a range of sites in the south of England and this study is reported in detail in Chapter 6.

A survey of beech health at 72 sites of conservation interest in southern Britain was undertaken by Imperial College in 1987 and 1988. Of the 1728 trees assessed, 23% had a loss of crown density of more than 25%. A similar proportion of trees were placed in categories 2 and 3 using the Roloff assessment, while 5% had moderate or severe crown chlorosis. Large differences in health were found between adjacent sites, suggesting that local site factors were of major importance in influencing tree health.

Statistical analysis identified some factors which explained the variation between and within sites. For the Roloff assessment, but not crown density, older trees were in worse health. Stand density strongly affected the crown density assessment, with thinner crowns being found in the more open stands as in the Forestry Commission survey. Local disturbance, in the form of old felling or windblow, affected both crown density and the Roloff assessment. Soil conditions were also found to be significantly related to tree health. Chlorosis, as expected, was largely confined to highly calcareous soils. However, both the crown density and Roloff assessments suggested that health was best on brown earths and sands and worst on acidic, poorly drained soils or on thin calcareous soils.

The survey indicated that trees from sites of high conservation value were in relatively poor health, whereas those in Forestry Commission ownership were in relatively good health. A likely explanation is that Sites of Special Scientific Interest tend not to occur on well-drained soils of moderate pH, but on the more acidic poorly drained soils or on thin calcareous soils.

Having identified local site factors of significance for beech health, the effects of adding data on pollutant levels or climatic variables within the survey area were investigated. For crown density, tree health was worse at sites with high SO_2 concentrations, and in regions which experienced the most severe drought in 1976. No clear relationships were identified for the Roloff assessments. Chlorosis was also worse in regions with the most severe drought in 1976, but the level of chlorosis was negatively related to the amount of nitrogen deposition.

It has been suggested that the impact of acidic or nitrogen deposition on forest health will depend on soil acidity. Accordingly, statistical analysis of relationships with pollution was repeated with the sites grouped into three pH ranges. It was found that the negative relationships between pollutant deposition and tree health were only found for acidic sites. In contrast, positive relationships between tree health and NH_3/NH_4 deposition were found on calcareous sites.

Sixteen of the 72 sites have been selected for a longer term monitoring programme. The results of this programme show evidence of a decline in beech health, especially over the past two years, in contrast to the trend shown by the main Forestry Commission survey. Of these trees 31.5% showed a loss of crown density of 25% or more in 1990 compared with 19.2% in 1987/88, while 44.4% were placed in categories 2 and 3 using the Roloff assessment, compared with 29.8% in 1987 (Power and Ashmore, 1991). A further decline in health is evident in the data for 1991.

This change in health is not consistent across the study sites. There is little evidence of change at sites on well-drained brown earths and sands, while the decline in health is more marked on sites on stagnogleys or rendzinas. This suggests that the high soil water deficits in recent summers may have been an important factor on the observed decline. At some sites, storm damage or local disturbance may have contributed to a decline in crown density, but at other sites this is not the case.

At six of the study sites, healthy and unhealthy trees have been compared in order to elucidate the possible causes of reduced vitality (Power and Ashmore, 1991). There was little evidence of differences between healthy and unhealthy trees in leaf nutrient content, but marked differences between them were found in both soil chemistry and root vitality. Soil aluminium concentration and aluminium/calcium ratios were consistently higher under unhealthy trees at three acidic sites, while soil potassium content was significantly higher under healthy trees at three sites. Healthy trees also had a higher proportion of live or mycorrhizal root tips at all but one of the sites. Historical patterns of extension growth also showed differences between the two groups of trees. At two sites, there was evidence that growth patterns only diverged after the drought period of the mid-1970s. The currently unhealthy trees did not recover to pre-drought growth rates, whereas the currently healthy trees did. However, at other sites, growth differences between the healthy and unhealthy trees were already apparent before 1975/6.

The survey by Tickle (1988) covered a similar geographical and age range, and type of site to the Imperial College survey. However, a rather greater number of trees (39%) had a loss of crown density of 25% or more, than in the Imperial College survey. The assessment of the percentage of trees in the Roloff categories 2 and 3 (48%) was also higher than that found by the Imperial College group, while nearly one third of the trees had some chlorosis. This latter observation, however, is of dubious value as the survey was carried out in late August and early September.

Ash

Ash is not particularly important as a forest species, constituting only 4% of the forest area (Locke, 1987). However, it has been widely planted in hedgerows and, following the loss of elms throughout much of Great Britain, it forms an important component of the landscape. It has not featured in any specific survey, although a specific study of dieback in hedgerow ash has been carried out and this is described in Chapter 6. In the annual surveys conducted by the Forestry Commission on behalf of the European Community, 79 ash trees are assessed. This sample is too small to draw any conclusions about the condition of the trees in Britain.

Yew

Yew is not a species that is generally grown commercially, although there has been a recent increase in interest in the possibility of planting this species in some lowland areas. At present, it is found mainly as an ornamental, although it occurs naturally in some types of woodland. It is susceptible to severe winter frosts which restricts its natural range towards the north of the country and it is also strongly hygrophilous. Consequently, it is susceptible to drought where it occurs on soils that dry out quickly.

The health of yew at 40 sites (320 trees) in southern England was assessed by Tickle (1988); 74% of the trees had a reduction in crown density of 25% or more. The presence of adventitious shoots and dying sub-branches was also reported to be common. Many of the trees were in churchyards or at roadsides, where local stress and disturbance is likely, and their age was unknown. Thus the significance of this high percentage of trees with some loss of crown density is difficult to evaluate.

Other species – United Nations Economic Commission for Europe (UN-ECE) Surveys

As part of the international survey run by the Commission of the European Communities, the Forestry Commission monitors the crown condition of trees at 71 sites in Great Britain. The sites are distributed on a 16 km by 16 km grid, with the majority of sites being located in southern England or Scotland. At each site, 24 trees are assessed and, in contrast to the main monitoring programme undertaken by the Forestry Commission, there is no restriction on species. The numbers of trees of each species examined are presented in Table 3.5. Most (63%) of the trees are conifers and the majority (76%) are less than 60 years old. The survey, which is undertaken as a direct result of European Community legislation, has limited value and the results are difficult to interpret because of the small sample sizes involved and the nature and distribution of the sample plots (Innes, 1988b).

Despite these difficulties, the results of the most recent UN-ECE survey of tree health in 1991 cannot be discounted (UN-ECE, 1992). The condition of UK trees is shown to be generally poorer than in other European countries. Of the trees observed in the survey, 58% were shown to be moderately or severely defoliated compared with 25% in Germany and 7% in France (Table 3.6).

It is accepted that at many of the survey sites the trees are in poor condition. There are several contributing factors. Insect pest attacks and climatic effects including summer drought, winter frost and wind damage have made major contributions to these results.

The survey results for individual countries also produce differences as a result of differences in protocols for assessment of tree conditions in different countries. Studies by Landmann (1990) of protocol differences between national survey teams, show for example that defoliation scoring of the same trees by UK and French teams differed by 12%, the UK recording 12% more foliar loss. Similarly, scoring by the UK and Germany differed by 6%, the UK team recording 6% more foliar loss. These differences in assessment techniques, do not however account for the generally poorer tree health observed in UK trees.

The 1991 survey also shows that tree health in terms of defoliation in the UK has declined since 1987 when UN-ECE surveys commenced with a particularly dramatic decline in health of

Table 3.5. Numbers of trees sampled in the British portion of the UN-ECE and Commission of the European Communities surveys of forest health (from Innes 1990).

Species	<60 years old	>60 years old	Mixed Age Stands	Total
Acer pseudoplatanus	15	15	32	62
Alnus glutinosa	39	1	1	42
Alnus viridis	0	4	0	4
Betula pendula	26	2	9	37
Betula pubescens	31	6	25	62
Carpinus betulus	20	0	4	24
Castanea sativa	30	0	4	34
Corylus avellana	0	0	8	8
Eucalyptus sp.	4	0	0	4
Fagus sylvatica	17	19	36	72
Fraxinus excelsior	30	31	18	79
Ilex aquifolium	0	0	6	6
Populus nigra	13	0	0	13
Populus tremula	1	0	0	1
Prunus avium	2	2	2	6
Quercus petraea	2	24	2	28
Quercus robur	31	55	50	236
Quercus rubra	2	0	0	2
Salix sp.	0	0	1	1
Sorbus aucuparia	5	4	1	10
Tilia cordata	0	2	0	2
Ulmus glabra	2	0	0	2
Ulmus minor	0	0	9	9
Other broadleaves	28	2	6	36
Abies alba	1	0	0	1
Larix decidua	28	0	0	28
Larix kaempferi	63	0	3	66
Picea abies	48	0	0	48
Picea sitchensis	507	0	7	514
Pinus contorta	126	0	0	126
Pinus nigra	31	0	0	31
Pinus radiata	0	0	6	6
Pinus sylvestris	210	0	26	236
Pseudotsuga menziesii	62	1	8	71
Thuya sp.	3	0	0	3
Total	**1380**	**168**	**264**	**1812**

broadleaved trees between 1990 and 1991. These data highlight the need to follow up the broad trends with studies to quantify the role of insect pests and pathogens, climatic factors and pollutant factors at sites and in areas where damage is severe.

Interpretation of Survey Results

The results of the various different surveys can best be assessed by addressing a number of questions.

How does tree health in Great Britain compare with tree health in Europe?

The data for the United Kingdom are frequently compared with those from other countries. Such comparisons are very difficult to interpret because of the differences in the standards used for assessing trees. For example, Forestry Commission surveyors assess crown density more pessimistically than surveyors in France, Germany and the Netherlands. Another problem is that the sample sizes involved in the Forestry Commission's systematic, 16 by 16km grid survey of forest condition (which is the survey that should be directly comparable with continental surveys) are very small, with many species only being represented by a single tree (UN-ECE 1988, 1990, Commission of the European Communities 1989). Consequently, data obtained from surveys of tree health in Britain and elsewhere in Europe were not always comparable. Despite differences in assessment protocols, results of the most recent UN-ECE surveys indicate that tree health in the UK is generally poorer than in other European countries.

Is there any indication of a decline in tree health in Britain?

The surveys have not been conducted for a sufficiently long period to indicate whether any long-term trend in crown density is occurring. It is thought that at least ten years of data will be required before any such trend can be identified.

Studies of twig extension from felled beech trees in southern Britain have provided evidence of a longer-term decline in beech growth at some sites since the 1970's (Lonsdale et al., 1989). This study is discussed in more detail in Chapter 6. More recently work by Power and Ashmore (1991) has shown that trees which are currently in poor condition began to decline in extension growth, compared with trees currently in good condition at the same site, at least 15 years ago. Thus data collected in surveys may reflect long-term changes in health occurring over periods of more than a decade.

It is possible that changes of twig growth in beech identified using Roloff's assessment methods indicate an acceleration in the normal pattern of growth decline that occurs as a tree ages, but this interpretation is being increasingly questioned. The system adopted by Roloff for the assessment of recent twig growth has a number of important limitations, ranging from difficulties of application and reproducibility (Innes and Boswell, 1990b) to problems with the interpretation of the results (Thiebaut, 1988; Athari and Kramer, 1989). The nature and extent of crown dieback and crown structure appear to offer more reliable measures of long-term effects, but more work is required on these before any substantive conclusions can be reached.

Is tree health in Britain good or bad?

A critical question is whether or not tree health is worse than it might be expected to be. To establish this, it is necessary either to compare the condition of trees in Britain with those elsewhere or to compare the condition of trees with their condition in the past. Valid international comparisons of survey data are difficult to make and, the differing growth conditions in Britain make such comparisons very difficult to interpret. However evidence from the most recent 1991 UN-ECE survey suggests that tree health in the UK is poorer than in other European countries. The evidence for

24

Table 3.6: Defoliation (classes 2-4) in conifers and broadleaves in UN-ECE tree health surveys in 1986–1991. (Taken from: Forest Condition in Europe, 1992).

a)	Conifers						
			Defoliation classes 2–4				% change
	1986	1987	1988	1989	1990	1991	1990/91
Austria	—	—	12.0	10.1	8.3	7.0	–1.3
Belgium	—	4.7	10.8	15.0	26.5	23.4	–3.1
Bulgaria	4.7	3.8	7.6	32.9	37.4		
Byelorussia	—	—	—	76.0	57.0		
Croatia							
Czechoslovakia	16.4	15.6	27.0	32.0	50.3	46.0	–4.3
Denmark	—	24.0	21.0	24.0	18.8	31.4	12.6
Estonia	—	—	9.0	28.5	20.0	28.0	8.0
Finland	—	13.5	17.0	18.7	18.0	17.2	–1.6
France	12.5	12.0	9.1	7.2	6.6	6.7	–0.1
Germany (a)	19.5	15.9	14.0	13.2	15.0	24.8	9.8
Greece	—	—	7.7	6.7	10.0	7.2	–2.8
Hungary	—	—	9.4	13.3	23.3	17.8	–5.5
Ireland	—	—	4.8	13.3	5.4	15.0	9.6
Italy	—	—	—	—	—	13.8	
Latvia	—	—	—	—	43.0		
Lithuania	—	14.8	3.0	24.0	22.9	27.8	4.9
Liechtenstein	22.0	27.0	23.0	12.4			
Luxembourg	4.2	3.8	11.1	9.5			
Netherlands	28.9	18.7	14.5	17.7	21.4	21.4	0.0
Norway	—	—	20.8	14.8	17.1	19.0	1.9
Poland	—	—	24.2	34.5	40.7	46.9	6.2
Portugal	—	—	1.7	9.8	25.7	19.8	–5.9
Romania	—	—	—	—	—	6.9	
Russia	—	—	—	—	—	4.2	
Slovenia	—	—	—	—	34.6	31.3	–3.3
Spain	18.2	10.7	7.3	3.5	3.1	7.3	4.2
Sweden	11.1	5.6	12.3	12.9	16.1	12.3	–3.8
Switzerland	16.0	14.0	15.0	14.0	19.0	21.0	2.0
Turkey							
Ukraine	—	—	—	1.4	3.0		
United Kingdom	—	23.0	27.0	34.0	45.0	51.5	6.5
Yugoslavia (b)	23.0	16.1	17.5	39.1	34.6	15.9	–18.7

a) 16 x 16 km network after 1988   (a) East and West   (b) Croatia, Slovenia excluded

Table 3.6: Defoliation (classes 2–4) in conifers and broadleaves in UN-ECE tree health surveys in 1986–1991. (Taken from: Forest Condition in Europe, 1992).

b) Broadleaves

	1986	1987	1988	1989	1990	1991	% change 1990/91
Austria	—	—	16.6	15.7	14.9	11.1	–3.8
Belgium	—	16.0	10.0	8.1	10.1	13.5	3.4
Bulgaria	4.0	3.1	8.8	16.2	17.3		
Byelorussia	—	—	—	33.4	45.0		
Croatia							
Czechoslovakia	—	—	29.1	37.0	33.9	23.7	–10.2
Denmark	—	20.0	14.0	30.0	25.4	27.3	1.9
Estonia		Only conifers assessed					
Finland	—	4.7	7.9	12.6	11.6	7.7	–3.9
France	4.8	6.5	5.3	4.8	7.7	7.4	–0.3
Germany (a)	16.8	19.2	16.5	20.4	23.8	26.5	2.7
Greece	—	—	28.5	18.4	26.5	28.5	2.0
Hungary	—	—	7.0	12.5	21.5	19.9	–1.6
Ireland		Only conifers assessed					
Italy	—	3.6	2.9	9.5	16.7	17.1	0.4
Latvia	—	—	—	—	27.0		
Lithuania	—	—	1.0	16.0	15.8	14.9	–0.9
Liechtenstein	10.0	7.0	5.0	9.0			
Luxembourg	5.6	10.1	12.3	13.9	—	33.9	
Netherlands	13.2	26.5	25.4	13.1	11.1		
Norway	—	—	—	—	18.2	25.1	6.9
Poland	—	—	7.1	17.7	25.6	34.8	9.2
Portugal	—	—	0.8	8.6	34.1	36.6	2.5
Romania	—	—	—	—	—	10.4	
Russia	—	Only conifers assessed					
Slovenia	—	—	—	—	4.4	5.8	1.4
Spain	13.7	13.7	6.8	3.2	4.4	7.4	3.0
Sweden	—	—	5.2	—	22.1	9.1	
Switzerland	8.0	15.0	7.0	6.0	12.0	13.0	1.0
Turkey							
Ukraine	—	—	—	1.4	2.7		
United Kingdom	—	20.0	20.0	21.0	28.8	65.6	36.8
Yugoslavia (b)	—	7.3	9.0	8.2	4.4	8.2	3.8

a) 16 x 16 km network after 1988   (a) East and West   (b) Croatia, Slovenia excluded

changes through time is controversial although a considerable amount of work is currently being done in this area. Again evidence from the recent 1991 UN-ECE survey suggests that tree health has declined in the UK particularly over the past few years. While it is possible to determine the state of health of trees at a particular site, where comparisons can be made with neighbouring sites, it is more difficult to evaluate the state of health of trees on a national basis.

What instances are there where a cause can be confidently assigned to poor tree condition?

The annual surveys of forest condition have revealed marked year-to-year variations in the condition of individual trees. In many cases, the primary cause of these variations can be readily identified. For example, the storms of 1987 and 1990 resulted in severe structural damage to some trees, which was recorded in the surveys. Similarly, the droughts of 1976, 1989 and 1990 have all been associated with a deterioration in the health of a number of species, particularly beech. These years also had more frequent ozone episodes than average. For both storm damage and drought-induced mortality, the opening up of stands following tree death may have contributed to the deterioration of the remaining trees, but this is rather less certain.

Damage by insects and fungal attack can also be confidently identified. In some cases, such as with Sitka spruce in 1989, the defoliation may be widespread and severe. In such cases, the primary cause is clearly the defoliating agent. However, the direct or indirect impact of pollution may also be involved (see Chapter 5), with pollution-stressed trees possibly being more susceptible to insects or fungi or pollution itself possibly having a direct effect on the organisms. Clearly, there are many instances where poor tree condition can be explained without invoking any pollution-related mechanism.

Under what circumstances are the causes of poor crown condition impossible to determine?

One of the basic products of surveys is information on the spatial distribution of specific phenomena. For instance, the surveys of crown condition produce information on the condition of trees at a wide variety of locations. These locations experience very different environmental conditions, including pollution climates. It has been suggested that spatial trends in tree condition may correspond with trends in pollution climate.

This hypothesis is fundamental to several of the approaches to the interpretation of survey data that have been used in the past. However, it has a major drawback. Although there are clear trends in the distributions of specific pollutants, these trends are frequently correlated with those of other environmental parameters. Gaseous pollution tends to be higher in the south and east of the country, but this area also has warmer and drier conditions than elsewhere. The association is particularly marked in the case of ozone and drought on a regional scale although local soil conditions will modify the level of drought stress experienced by the trees. Consequently, it is frequently impossible to separate the effects of pollution from the effects of other environmental variables. This problem has dominated attempts to interpret British survey data in relation to pollution and has yet to be satisfactorily resolved. Local site conditions will also modify pollutant impact and directly influence tree health, and need to be incorporated in future efforts to identify relationships between tree health and environmental conditions.

Can the influence of air pollution on trees be identified from the survey data?

The problems described above make it very difficult to relate survey data to pollution. This does not mean that survey data have no value. Survey data can be used to identify problems and, if sufficiently detailed, can help generate cause-effect hypotheses which can later be tested by appropriate methods. To undertake such analyses, it is essential to have good information on the pollution climate at each site. It is now clear that the pollution climate within a forest stand may differ significantly from the pollution climate at an immediately adjacent open site and this has major implications for the type of analysis by association that has been undertaken in the past. There is a clear need for much more information on pollution climates within forest stands before extroplations from existing monitoring networks can be reliably undertaken.

A major problem that is evident from survey data is the non-specific nature of the symptoms that have been assessed. Whilst there have been significant developments in the characterisation of tree condition, the lack of specific symptoms that can be directly related to the influence of pollution remains a major drawback.

The Early Diagnosis Study by the Institute of Terrestrial Ecology

In an attempt to quantify the influence of air pollution on trees, an early diagnosis study was undertaken by ITE at four sites in Britain (Cape et al., 1988).

The four British sites formed part of a wider network of twelve sites spread across an area extending from southern Germany to the north of Scotland. Scots pine, Norway spruce and beech were studied with up to 12 trees of any given species being sampled at each site, resulting in a total of 250 trees being sampled. A range of different parameters were examined, ranging from leaf wettability to needle nutritional status.

Of the variables examined, a number were recommended for further investigation using experimental techniques. These were: amounts of surface wax, contact angles, the Härtel turbidity test, modulated fluorescence, hydrocarbon emissions, buffer capacity, pigments and alpha-tocopherol.

The authors of the report have drawn attention to the limitations of the survey. The most important of these is that it is impossible to identify cause-effect relationships on the basis of the survey. For example, differences between sites could be attributable to any of a range of environmental variables and the number of sites was insufficient to identify the most important factors. A further problem is the very small sample sizes involved. Most of the parameters that were investigated vary between trees and the adequacy of the sampling design at individual sites is unknown.

One of the measurements assessed by the Institute of Terrestrial Ecology was the nutrient content of foliage. Innes and Boswell (1989b) examined the sulphur contents of foliage from Sitka spruce, Norway spruce and Scots pine from throughout the UK and found that these were highly correlated with atmospheric sulphur dioxide concentrations, a finding also reported by Farrar et al., (1977) for an area in the Pennines in the 1970s. The foliar sulphur contents were unrelated to any of the indices of tree condition used in the survey. A relationship has also been identified between the nitrogen content of foliage and atmospheric concentrations of various nitrogen compounds (Forestry Commission, unpublished data). In contrast to results from West Germany and elsewhere, no magnesium deficiency has been reported (Binns et al. 1986).

4

Damage Caused by Factors other than Air Pollution

Conclusions

● As long-lived organisms, trees suffer from many damaging agents. Some agents are catastrophic e.g. gales, others are chronic e.g. aphids. Some are important on young trees e.g. voles, while some are important on old trees e.g. decay fungi.

● In the UK, the influence of man is often adverse, through destructive management practices in woodland and the provision of hostile environments in towns.

● Certain characteristics of the climate such as windiness and unseasonal frosts are common causes of damage to trees in the UK.

● Because tree planting in the UK includes a large proportion of exotic species, or of exotic provenances of native species, some at least of our tree population is likely to be poorly adapted to aspects of our climate, such as unseasonal frost.

● From time to time exotic organisms capable of damaging trees are likely to reach the UK, e.g. Dutch elm disease and the great spruce bark beetle.

● In consequence it is <u>normal</u> for a proportion of trees to be showing some evidence of past or recent damage which would be reflected in crown condition and general appearance.

Introduction

During its life a tree is subject to a great many adverse influences. These may result in prolonged periods of poor or distorted growth or even death. Alternatively, the effects may be short-lived and rapid recovery may occur. Young trees are generally very vulnerable to damage, and many factors of great importance at this stage in the tree's life may be of little or no significance subsequently.

This chapter briefly reviews some principal causes of death and damage to trees in Britain. Although air pollutants are not considered here, there is increasing evidence that air pollutants may alter the sensitivity of trees to some of the factors reviewed in this chapter. Interactions are discussed in Chapter 5. The chapter starts with a consideration of the activities of man himself, excluding his role as a producer of air pollutants, and then covers natural abiotic damage and biotic damage. Finally reference is made to damage resulting from the combined action of two or more agents. Most of the examples involve damaging events occurring within the last 20 years.

The Influence of Man

The woodlands of Britain have been disturbed and manipulated by man for thousands of years by felling, coppicing and planting. During this time he has introduced many new tree species and has also brought in foreign seed sources of native species. Poor growth may sometimes indicate an unsuitable choice of species for a site, or some adverse change in the woodland environment. Beech, for example, shows crown deterioration if the stand is subject to heavy thinning or if adjacent blocks of woodland are cleared. The condition known as top dying of Norway spruce is exacerbated by the same kinds of practices. Many young trees which are transplanted from the nursery to the forest die because of poor site preparation or

Plate 3.1   Crown Density in Scots pine – 0% loss. (From Innes, 1990).

Plate 3.2   Crown Density in Scots pine – 40% loss.
(From Innes, 1990).

Plate 3.3   Crown Density in oak – 10% loss. (From Innes, 1990).

Plate 3.4   Crown Density in oak – 55% loss.
            (From Innes, 1990).

Plate 4.1   Spring frost damage to Sitka spruce.

Plate 6.1   Damage to Norway spruce downwind of an intensive animal
             rearing unit in North Yorkshire.

Plate 6.4   Needle browning and shoot death on 58-year-old Scots pine (in mixture with Norway spruce)
            in Thornwaite Forest, near Keswick, Cumbria, August 1984. Elevation 350–400 m.

Plate 6.2   Foliar damage to hawthorn in Lincolnshire –
            October 1989.

Plate 6.3   Severe ash dieback in a tree now isolated in an arable
            field though located on the line of an old hedgerow.

poor plant handling (Nelson, 1990). Many also die subsequently because of poor nutrition or drainage, but most of all because of weed competition (Potter, 1989).

In towns, poor planting, establishment and maintenance practices result in the death of thousands of trees annually. Thus Gilbertson and Bradshaw (1990) reported that in inner city Liverpool 39% of trees died within 5 years of planting. With established trees, much damage results from soil compaction, root disturbance during construction work and water-logging through changes in drainage. De-icing salt can cause serious injury (Dobson, 1991), as can inappropriately applied herbicides, gas leaks and vandalism. Mechanical wounding and pruning can provide sites for pathogen entry (Lonsdale, 1989).

Many of these problems also occur in the countryside, especially near major roads. In addition there are particular problems that are associated with agricultural practices such as the severance of tree roots during cultivation, herbicide spray drift and stubble burning. Trees planted on reclaimed land sometimes suffer from poor drainage and metal toxicity.

Natural Abiotic Damage

Climatic factors

In general, the climate of the UK is favourable for the growth of trees, and the climax vegetation is mixed deciduous woodlands in the lowlands and Scots pine/birch forest in much of the uplands. Many exotic species are also well suited to UK conditions. However, trees grown anywhere in the world are likely to encounter adverse climatic events at some stage during their long lifetimes. In the UK, these events include high winds, soil water deficits and unseasonal frosts. There are also various kinds of winter-related injury – some linked to very cold winters, others to mild ones. In addition there is the special case of fire.

Wind

High winds are by far the most serious agents of climatic damage to trees in the UK. Severe storms quite frequently cause catastrophic windthrow affecting thousands of hectares of woodland. During the last 50 years, these storms have occurred equally frequently in the north and the south of the country, and commonly affect areas of 300–700 km^2 where maximum gusts exceed 150 km h^{-1}. Table 4.1 from Quine (1991) lists the main storms of this kind that have occurred in the UK since 1945. In addition, some windthrow occurs every winter during gales with gusts of 60–110 km h^{-1}. This type of windthrow is most serious in the conifer plantations that have been established in the uplands where soil conditions frequently restrict root development. The loss of trees attributable to this kind of windthrow is, in fact, much greater than that resulting from catastrophic gales, with the area affected each year being around 1,100–2,200 ha, with an annual loss of 0.3–0.6 million m^3 timber.

Strong winds have important effects on the appearance of trees and, in conifers, on needle retention. In addition, wind-borne (and fog-suspended) marine salt can cause foliar injury many tens of kilometres inland, although few chloride measurements have been taken to determine the role of the salt compared with that of the wind itself.

Water stress

While the UK is often thought of as having plentiful and well-distributed rainfall, the annual average potential soil water deficit exceeds 100mm in most of lowland England (Green, 1964). Consequently, even in non-drought years, soil water availability can be a major constraint on tree growth in this part of the UK, especially where soils are shallow or have small water-holding capacities. Soil water deficits can also restrict tree growth in the uplands, because little water is retained in many of the shallow soils and also because fine roots die at surprisingly small water deficits in organic soils (Deans, 1979).

Table 4.1:   Summary of data for catastrophic storms affecting Britain since 1945 (from Quine, 1991).

Date of storm	Area affected by 130 km h⁻¹ gusts (km³)	Max. gust recorded (km h⁻¹)	Mean of max. gusts recorded within 130 km h⁻¹ zone	Volume of windthrown timber (m³ x 10⁶)	Growing stock windthrown (%)
31 January 1953	370	182	156	1.80	9.7–25.3
15 January 1968	510	189	154	1.64	15–30*#
2 January 1976	890	167	141	0.96	<5
16 October 1987	220	186	148	3.91*	13–24
25 January 1990	690	173	141	1.26*	1–3

*   Known to include non-woodland trees.　　　　*#   Percentage of crops aged 31 years and over.

In very dry summers visible tree damage occurs, taking the form of leaf discolouration or premature leaf fall. During the last 20 years this has been most noticeable in 1976, 1984, 1989 and 1990. Shallow-rooted species like birch are the most seriously affected but in years like 1989 and 1990 even normally deep-rooted species like oak are affected on certain soils. The consequences of severe drought can continue into succeeding years with many species showing very poor shoot and girth growth in the years after a drought *(for further information on this point see Chapter 6)*.

Frost damage

The shoots (apices, foliage and cambium) of most trees are susceptible to frost damage when air temperatures fall below about –2.5°C during the period of active growth. In the UK, most frost damage (Plate 4.1) occurs in spring, because growth is promoted by warm temperatures in March–April while there is a high probability of –2.5°C frosts in April–May, especially at high altitudes. The importance of spring frost damage to trees was brought to

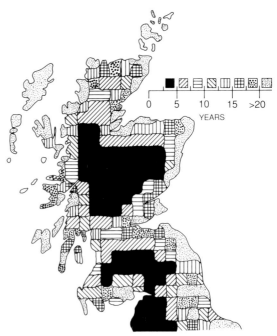

4.1 Return times of potentially damaging frosts (below –2.5°C), occurring near the date of budburst of young trees of Sitka spruce in northern Britain. (Taken from Cannell and Smith, 1984.)

the fore following severe frosts in mid-May 1935 and 1945, which caused widespread damage to trees and other vegetation across the UK (Day and Peace, 1946). More recently spring frost damage to trees has warranted mention in the Forestry Commission Annual Reports approximately every second year. Severe damage to fruit trees has been reported in parts of the UK in each of the past three years 1989–91. Oak, beech and ash are among the most susceptible of the native broadleaved species. Scots pine and birch are resistant. Non-native trees such as Sitka spruce are often affected, species showing death of both new shoots and sometimes main stems (Redfern, 1982). The risk of damage at the time of bud burst on young trees of this species was estimated by Cannell and Smith (1984) using a model to determine the date of bud burst and the probability of air frosts of $-2.5°C$ (Fig. 4.1). They estimated that in most of the upland plantation regions of Scotland, young trees of Sitka spruce are likely to suffer spring frost damage once every 3–5 years.

Frost damage to trees in the autumn has been less commonly recorded, partly because the symptoms are less well known, but also because most species become frost hardy before damaging frosts occur (Cannell, 1985). However, autumn frost damage in Sitka spruce has been observed frequently in nurseries and occasionally in young plantations in Scotland (Redfern and Cannell, 1982; Redfern et al., 1987).

Winter injury

Various kinds of winter-related injury occur when trees are not actively growing. In extremely cold winters, such as that of 1981/82, dieback can occur in both native species like pedunculate oak (*Quercus robur*) as well as exotics like *Nothofagus* spp. and Leyland cypress (*Cupressocyparis leylandii*) (Strouts and Patch, 1983). A striking feature of this winter was the killing of bark that occurred on the main stem of conifers such as Corsican pine (*Pinus nigra*), with crown symptoms sometimes being delayed for up to two years (Redfern and

Rose, 1984). Desiccation damage to conifers is also associated with winter conditions in which cold spells alternate with warm windy weather. Top dying of Norway spruce, a common disorder in the drier parts of Britain, is linked with mild windy winters (Diamandis, 1979a,b).

Fire

This is a rather special cause of damage, since in Britain virtually all fires are started by man. Weather conditions have, however, a profound influence on the extent of the losses, with the largest areas of woodland destroyed in the last 20 years occurring in the hot dry summers of 1976, 1984, 1989 and 1990. Fire-killed trees remain conspicuous long after the surrounding herbaceous and shrub vegetation has regrown. As indicated earlier, damage to hedgerow trees can often follow poorly controlled stubble burning.

Lightning

Lightning is a common cause of damage with groups of trees up to 50m across being killed on occasions.

Nutrient deficiencies

Trees, in common with other plants, need a balanced supply of a range of nutrient elements for healthy growth. The macro-nutrients, nitrogen, phosphorus, potassium, magnesium, calcium and sulphur are required in relatively large amounts while the micro-nutrients copper, iron, boron, manganese are only required in trace amounts. A deficiency of any nutrient can influence the health of trees and, if acute, can produce characteristic visible symptoms such as discolouration of foliage or change in growth form (eg Binns et al., 1980). Deficiencies occur naturally as a result of inherent differences in soil properties or due to processes such as acidification. They may arise due to low levels of a particular nutrient in the soil or because

32

a nutrient, although present in the soil, is not available in a form which can be taken up by the trees. The main areas in which particular deficiencies occur in the UK are well known. For example, a deficiency of potassium is relatively common in plantation forests on deep peat in the uplands, while phosphorus deficiency is common in first rotation plantations on a range of upland soils. Iron deficiency can induce chlorosis on calcareous soils because the iron does not enter solution at the high pH of these soils. Deficiencies of one element may be induced if other elements, are available in plentiful amounts and growth is rapid. Thus, a temporary copper deficiency can occur during early growth. Induced deficiencies can also arise as a result of adverse soil physical conditions such as compaction, induration or poor drainage.

Nutrient deficiencies can generally be overcome by the application of fertilizers or in the case of nitrogen deficiency on heather ground by suppressing competing ground flora. However, where there are poor soil physical conditions, fertilizer applications may not overcome nutrient deficiencies.

Metal toxicity

A number of heavy metals, eg copper, zinc, lead, cadmium can cause growth deformation and/or poor growth when present in greater than trace concentrations. Naturally occurring toxic levels of trace metals are rare and are associated with localised geochemical anomalies (eg part of the Coed y Brenin forest, north Wales). Metal concentrations in serpentinite rocks can also sometimes be high enough to cause reductions in growth but, as with geochemical anomalies, the distribution of these rock types is well known.

Biotic Damage

Conspicuous symptoms in the crowns of established trees can be caused by a great variety of living agents including bacteria, viruses, nematodes, etc. However, the three most important groups of organisms associated with such damage are fungi, insects and mammals.

Fungal diseases

During the period 1970–90, the most serious pathological event for trees in the UK has undoubtedly been the Dutch elm disease epidemic. At its height in the early 1970's, surveys in southern Britain showed that trees were dying at a rate almost unequalled in the annals of plant disease (see Fig. 4.2) and by 1983 it was estimated that about 85% of the 23 million elms in this part of the country had been killed (Gibbs, 1978; Greig and Gibbs, 1983). The disease has developed rather more slowly in northern Britain, but even so a high proportion of the elms south of the Great Glen have been killed. New populations of elm are growing up rapidly in many areas but these must be expected to suffer outbreaks of the disease in

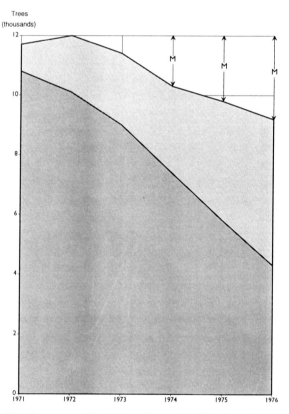

4.2 Progress of Dutch elm disease in 234 plots distributed through southern England. Dark stippling, trees healthy or only slightly affected: light stippling, trees dying or dead: M, net loss of diseased trees through felling.

their turn. It should be noted that the epidemic resulted from the introduction to Britain of a new form of the causal fungus now named *Ophiostoma novo-ulmi* (Brasier and Gibbs, 1973; Brasier, 1991). The disease does not develop rapidly in stressed trees – rather the reverse – although the hot, dry summer of 1976 did promote the activity of the vector bark beetle (Gibbs and Greig, 1977).

Another pathogen warranting special mention is the honey fungus (*Armillaria*). This fungus, which exists as a number of species each with different ecological and pathological attributes, can kill trees when they are growing in the vicinity of old stumps which provide the fungus with a food base. Under UK conditions, losses in woodland usually cease after the crops reach about 15 years of age, but the fungus is often a major cause of mortality in ornamental plantings, with the genera *Cedrus, Chamaecyparis, Malus, Prunus, Sequoiadendron* and *Sorbus* being among the most susceptible (Greig *et al.*, 1991).

Among other diseases causing dieback and death, mention should be made of *Brunchorstia* shoot dieback (*Brunchorstia pinea*) which occasionally affects young pole-stage plantations of Corsican pine in south-east Britain, and precludes the use of this species in more northerly and westerly areas (Gibbs, 1984). It has also been involved in the deterioration of some high elevation plantations of Scots pine in upland Britain (Redfern and Gregory, 1991). A rather similar dieback caused by the fungus *Ramichloridium pini*, is currently quite severe on certain ecotypes of lodgepole pine in Wales (Rose, 1991). The pine stem rust (*Peridermium pini*) is widespread on mature Scots pine in the Moray Firth area and is common in the large man-made Thetford Forest, East Anglia (Greig, 1987; Gibbs *et al.*, 1987). Also in Thetford the killing of pine by the root rot pathogen *Heterobasidion annosum* is a locally serious problem (Grieg, 1984).

Many fungi cause diseases of the leaves and current shoots of broadleaved trees. Some, such as the anthracnose diseases (eg *Marssonina salicicola* on willow and *Apignomonia errabunda* on London plane) can contribute to crown thinning and sometimes to dieback. Depending on the weather, trees may restore a full canopy by mid-summer. Other fungi, notably the rusts, can cause striking foliage discoloration. An example was provided by *Melampsoridium betulinum* on birch in northern UK in 1985 and 1987 (Redfern *et al.*, 1987a; Gregory *et al.*, 1988).

Many of the commonly planted conifers are largely free from serious leaf diseases, but in Scots pine needle browning and needle loss caused by the fungus *Lophodermium seditiosum* has been common and severe in some years in upland UK. Swiss needle cast (*Phaeocryptopus gaumannii*) sometimes causes considerable yellowing and defoliation of Douglas fir and was common in south-west England in 1989.

Insect attack

Death due to insect attack is very common on young trees on old woodland sites. The pine weevil (*Hylobius abietis*) is of special importance and can cause the death of a high proportion of newly planted trees (Heritage *et al.*, 1989). Bark beetles in the genus *Hylastes* are serious pests under similar conditions. The pine shoot beetle (*Tomicus piniperda*) can be an important factor in the death of Scots pine plantations weakened by other agents (see under disease complexes) and can cause considerable crown damage through shoot pruning. The great spruce bark beetle (*Dendroctonus micans*) became established in east Wales and parts of western England during the 1970s, although it was not discovered until 1982 (Fielding *et al.*, 1991; Gregoire, 1988). Over the next 2 years some 64,000 trees showing signs of beetle attack were felled. Since then silvicultural and biological control measures have reduced beetle numbers (Speight and Wainhouse, 1989) and populations are currently too low to have a significant effect on the appearance of spruce plantations.

Several defoliating insects have aroused attention in the last two decades. Outbreaks of the spruce sawfly (*Gilpinia hercyniae*), occurred on both Norway and Sitka spruce plantations in Wales

34

during the period 1970–76, but populations have remained at a low level since that time. The pine beauty moth (*Panolis flammea*) came to prominence in 1976 when it defoliated and killed 120 ha of lodgepole pine in the northern Scotland. Since then high populations of this insect have occurred every few years and have necessitated control by aerial applications of insecticides over some 20,000 ha. The pine looper moth (*Bupalus piniaria*) periodically causes severe defoliation, principally on Scots pine plantations in the English Midlands and parts of Scotland (Barbour, 1988). The insect population is monitored each year in forests known to be at risk. Outbreaks are treated with aerial applications of insecticide. Some 1,200 ha were treated in 1984.

One of the most damaging agents to the foliage of spruce is the green spruce aphid (*Elatobium abietinum*). Cycles of infestation can result in heavy loss of needles in successive years, leaving the tree in a very poor condition. Mild winters favour insect survival, but an air frost falling below –8°C is sufficient to cause a great reduction in the population and prevent serious defoliation of Sitka spruce (Carter, 1972). The effects of *Elatobium* infestation were particularly noticeable following mild winters in 1979, 1980 and 1989.

On oak, combined attacks by the oak leaf roller moth (*Tortrix viridana*) and the winter moth (*Operophtera brumata*) have led to conspicuous defoliation of large areas of trees in southern England during the late spring of some years – notably 1982 and 1983. Refoliation is normally complete by the end of June *(but see below under disease complexes)*.

Many other insects can have an effect on leaf appearance in broadleaves. A notable example is the beech leaf miner (*Rhynchaenus fagi*) which causes part of the leaf to turn brown. Damage caused by this insect is common in most years.

Mammal damage

During the period of tree establishment, a number of mammals can cause significant damage. Voles and rabbits can kill seedlings and young trees, while deer cause growth reduction and stem deformation through their browsing activities (Gill, 1991a,b). All three groups of mammal are distributed throughout the UK with the exception of deer which are absent from much of Wales.

The grey squirrel is a major pest on established trees. Crown dieback, following bark-stripping of the main stem, is common in sycamore and beech and is less common in oak and sweet chestnut. There is a close link between damage severity and squirrel population levels. Particularly bad years were 1972 and 1983.

Diseases with multiple causes

There are a number of problems that clearly involve several biotic and abiotic factors. One example is the dieback of upland Scots pine plantations which is attributed to the combined effects of the pine needle fungus (*Lophodermium seditiosium*), the pine shoot disease (*Brunchorstia pinea*) and the pine shoot beetle (*Tomicus piniperda*) (Redfern and Gregory, 1991). Another example of economic importance is beech bark disease which has caused appreciable mortality in pole-stage beech plantations in southern Britain (Lonsdale and Wainhouse, 1987). In this disease, infestation of the tree stems by the felted beech coccus (*Cryptococcus fagisuga*) is followed by bark-killing by the fungus *Nectria coccinea*. This disease was serious in pole-stage beech plantations in southern England during the 1970's, although losses have not prevented the development of a final crop. It is interesting to note that the drought of 1976 had a similar effect to the coccus in stressing trees and making them vulnerable to *N. coccinea* attack.

The causes of other tree diebacks are less clearly understood. An example is the periodic dieback of pedunculate oak (*Quercus robur*) in various parts of Britain. Defoliation by caterpillars is implicated, and honey fungus (*Armillaria* spp.) and other root-rotting fungi appear to play a part in some cases. Other factors also seem to be involved, however, and this problem is the subject of current investigation.

5

Air Pollution and Tree Health: Experimental Evidence

Conclusions

- Most of the experimental evidence is, and is likely to remain in the immediate future, based on relatively short-term experiments (ie < 5 years) on juvenile trees. Extrapolation from such experiments to longer-term effects of air pollutants on mature trees is uncertain.

- Filtration experiments have shown that exclusion of gaseous pollutants improved the growth of beech and Norway spruce at a site in southern England, but not at sites in the Pennines or in Scotland.

- Fumigation experiments have shown that realistic concentrations of ozone may reduce the growth of beech and Sitka spruce, but not Norway spruce. There is little evidence that current UK rural concentrations of SO_2 and/or NO_2 reduce the growth of major tree species. Experimental applications of acid mist equivalent to the largest deposition rates found in the UK uplands have been shown to reduce the growth of Sitka spruce.

- Realistic concentrations of gaseous pollutants and acid deposition can cause subtle changes in the morphology, physiology and biochemistry of trees. These may lead to alterations in the sensitivity of trees to other environmental stresses such as winter injury, water stress and attack by pests or pathogens. Pests and pathogens may also be directly affected by pollutants.

- In some cases the experimental data allow generalisations to be made about pollutant/stress interactions. For example, SO_2 and NO_2 are consistently more damaging at low temperatures, and aphid pests consistently perform better on trees fumigated with SO_2 and NO_2. In other cases the nature of the interaction is variable. For example, O_3 can both increase or decrease the performance of aphid pests and plant pathogens.

- Forest soils in parts of the UK have been acidified by the deposition of pollutants in addition to that normally associated with forest growth. Problems associated with soil acidification include nutrient imbalance, metal toxicity, changes in mycorrhizal development and species composition, and reduced rates of organic matter decomposition. Although these changes are likely to be detrimental to tree vitality, there is no direct evidence of soil acidification affecting canopy condition in the UK at the present time.

Introduction

There are obvious difficulties in establishing experiments which would allow the impact of air pollution to be assessed over the lifetime of a tree. Some work in the UK is currently aimed at developing experimental techniques which could be used for this purpose, but techniques are not at a stage where many full-scale field experiments on fully mature trees can be envisaged. Thus evidence must be drawn from shorter-term experiments (i.e. of periods of less than five years) on young trees (i.e. less than 10 years old). These can provide information on the potential impacts of current ambient pollutant concentrations on tree vitality or help to improve our understanding of the mechanisms by which pollutants may directly, or indirectly, influence tree health. However, there will always be uncertainties in using the results from such experiments to draw conclusions about the long-term impact of pollutant stress on forest ecosystems due to the differing response of young and mature trees to pollutant exposure.

Results from such experiments have already been reviewed by UK TERG (1988). This chapter aims to update that review, concentrating on recent research in the UK, which has provided new evidence on the impact of air pollutants, both alone and in combination with other major abiotic and biotic stresses.

Effects of Ambient Air Pollution

In 1987–1988 the Forestry Commission established experiments using open-top chambers to assess the effect of ambient air quality on the growth and physiology of young trees of four species of major economic or amenity value i.e. Sitka spruce, Norway spruce, Scots pine and beech. Chambers have been established at three locations (Headley in Hampshire; Chatsworth in Derbyshire and Glendevon in Perthshire) chosen to provide a range of pollution climates (Willson *et al.*, 1987). All four species are grown at Chatsworth, but beech is not grown at Glendevon and Sitka spruce is not grown at Headley because of the unsuitable climates. Each site has eight filtered air chambers, eight ambient air chambers and eight outside plots, but because it is not possible to grow every species in each chamber, there are effectively four replicate chambers per species and treatment at each site (Lee *et al.*, 1990a).

The experimental trees were planted directly into the soil at each site in March/April 1988, with the intention of continuing the experiment for five years. Published results from the study already have shown some evidence of significant effects of air filtration from a destructive harvest

carried out after 18 months, in November 1989 (Lee *et al.*, 1990b). No significant effects were found on fresh or dry weights, but the height of two species (Scots pine and Norway spruce) was significantly less in ambient air than in filtered air at Headley. There was no significant effect on height at the other two sites. Indeed, at Chatsworth (for Scots pine and Norway spruce) and at Glendevon (for Scots pine and Sitka spruce) there was evidence of a non-significant increase in tree height, of about 10%, in ambient air.

Data for 1990 show larger effects of treatment, but many of these effects are not statistically significant because of the high within-treatment variation (Durant *et al.*, 1991). The overall pattern is as for 1989, with significant decreases in growth in unfiltered air for some species and parameters at Headley, but no significant filtration effects at the other two sites. The significant effects were greatest on Norway spruce, for which stem and needle dry weights were 45% and 49% higher in filtered air than in unfiltered air. The stem diameter of Norway spruce and the height of beech were also significantly greater in filtered air, but no significant effects were found on Scots pine.

Other aspects of plant development and morphology have also been examined. The phenology of shoot development was monitored at weekly intervals during the spring of 1989 using a simple subjective assessment, but there was no evidence of any effect of ambient air pollution on shoot development of any of the species (Lee *et al.*, 1990b). Detailed studies of the size and morphology of the root system of beech have also been conducted at the Headley site, (Taylor and Davies, 1990). At the end of the 1988 season, root dry weight was decreased in ambient air, compared with filtered air, but root length was increased in ambient air, especially in the lower parts of the soil profile (Fig. 5.1). This tendency to produce longer, thinner roots may lead to greater damage to the roots during periods of water stress. Studies on Sitka spruce have also demonstrated an increased root length in ambient air at Headley, but in this case root weight was not significantly affected by treatment (Taylor *et al.*, 1989).

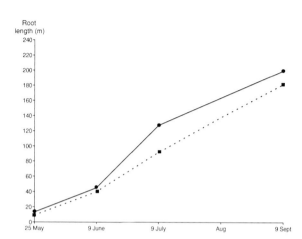

5.1 Growth of beech roots in filtered (■ — ■) and unfiltered (● — ●) air at Headley, Hamps., May–September 1988. Each point represents mean root length per plant, ± SE. May and June values were estimated non-destructively using gridded acrylic sheet. July and September values are for measurements of root length made destructively after separation of roots from soil. (From Taylor and Davies, 1990).

Physiological measurements have provided further evidence of an impact of air pollution at the Headley site. Taylor and Dobson (1989) reported that in the 1988 season, the stomatal conductance of first flush beech leaves was consistently lower in ambient air. On certain occasions when the difference between treatments exceeded 20%, the effect was statistically significant. In contrast, second flush leaves from these trees showed a non-significant increase in stomatal conductance in ambient air. Effects of filtration on the response of photosynthesis to CO_2 concentrations were also found in laboratory measurements on these trees.

The results from these experiments show clearly that ambient air pollution at the Headley site does influence tree physiology and morphology. However these effects are complex, and their significance for tree growth and vitality may depend on the prevalence of other stress factors. There is nevertheless evidence of reduced height growth in two species at this site. In contrast, at Chatsworth and Glendevon no significant effects of ambient air pollution have yet been detected.

Full details of the pollution levels at the three sites are not yet available, and this limits interpretation of data. However, the site at Headley is likely to experience relatively high O_3 concentrations, but lower concentrations of SO_2 and NO_x than the Chatsworth site. The possibility that O_3 is the primary pollutant responsible for the effects observed at Headley is supported by a recent open-top chamber fumigation study (Davidson et al., 1992). This shows that season-long exposure of young beech trees to O_3 concentrations within the range found in different summers in southern Britain significantly affected gas exchange and reduced root growth.

Interpretation of such chamber studies must always be tempered with caution because of the altered climate caused by the chambers. Comparison of outdoor trees with those inside the chambers at the Forestry Commission sites have shown earlier flushing of all species in the chambers (Lee et al., 1990b) and reduced dry weight and height growth in the outdoor trees in some cases (Lee et al., 1990a). These differences may be explained by the greater temperatures inside the chambers or the physical protection provided by the chambers. The extent to which these effects of chamber enclosure will become accentuated as the study proceeds is unclear, but the existence of some chamber effect does not negate the important data on the impact of ambient air pollution which have been generated from these experiments.

An open-air fumigation experiment was set up at Liphook, Hampshire by the then Central Electricity Research Laboratories in 1984 to investigate the effects of SO_2 and O_3 on tree growth and physiology of three species – Norway spruce, Sitka spruce and Scots pine, as well as other ecological processes. It uses a rather different technique from that adopted by the Forestry Commission. The trees are grown in the field and gases are dispersed across the plots from surrounding rings of pipework. This avoids the artefacts introduced by the use of chambers, but means that only the effects of pollutant concentrations above ambient can be examined. A further difficulty is that the size and cost of the system means that there is no replication of treatment plots.

a)

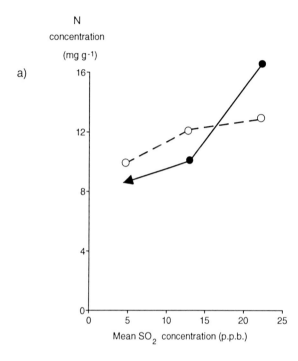

b)

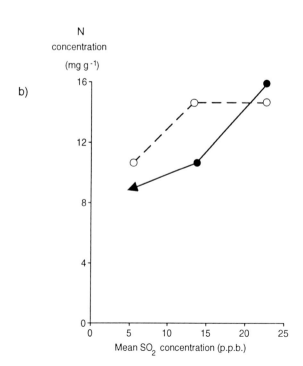

The trees were planted in 1985 and the experiment finished in 1991. SO₂ exposure started in March 1987 and O₃ exposure in May 1988. There were two SO₂ treatments (giving annual mean concentrations of 14 ppb and 24 ppb) based on monitoring data from the East Midlands. There was one O₃ treatment, at 1.5 times the ambient concentration at the site. In addition to visible injury and tree growth, studies of a range of other parameters (including gas exchange, mineral contents, carbohydrate and secondary metabolism, insect herbivore performance, litter decomposition, soil and water chemistry) have been undertaken on site. Thus the experiment is potentially of considerable importance in understanding the responses of young trees to SO₂ and O₃ over several years under realistic field conditions in southern Britain. The most important finding to date is the evidence of enhanced deposition of nitrogen in the plots receiving higher concentrations of SO₂, which is interpreted as being due to co-deposition of NH₃ (McLeod *et al.*, 1990). This was sufficient to substantially increase the needle nitrogen levels in Norway spruce (Fig. 5.2) and Sitka spruce in plots receiving the high SO₂ treatment, above that in the ambient plots which were below deficiency levels. The data suggest that co-deposition may be an important mechanism in increasing nitrogen inputs in areas of high SO₂ concentration, especially since the ambient NH₃ concentration at the site is not particularly high (annual mean 4 ppb).

These observations will obviously be very important in interpretation of any effects of SO₂ found at the end of the experiment. Interim results up to the end of the 1989 season show no effects on the growth of any of the three species, but some evidence of significant responses to SO₂. These preliminary data indicate that SO₂ increased

5.2 (left) Foliar concentrations of nitrogen in Norway spruce trees exposed to SO₂ and O₃ by open-air fumigation and sampled in October 1986. (a – N in 1987 needles; b – N in 1988 needles). Values are means of 8 trees for the 5 fumigated plots (○, ●), and 16 trees for the 2 unfumigated ambient plots (▲). Open symbols – plots receiving O₃ fumigation (30 ppb); filled symbols – ambient O₃ plots (24 ppb). Exposure concentrations are annual means for October 1987 to September 1988. (From McLeod *et al.* 1990.)

frost damage on Norway and Sitka spruce and accelerated needle loss from Scots pine as well as having other physiological effects.

The value of long-term experiments to investigate pollutant effects is further illustrated by some recent results from the outdoor closed-chamber fumigation facility at Lancaster University. Here 2-year old seedlings of Sitka spruce were exposed to ozone (70 ppb, 7 h mean) for a total of 285 days extending over 3 consecutive summers (1986–1988). Throughout this period and during 1989 when the trees were no longer exposed to the pollutant, the height and stem diameter of the trees were measured at regular intervals. Over the first and second years of exposure ozone had no effect on either the height or the diameter of the seedlings. Towards the end of the third summer of fumigation statistically significant reductions in both growth parameters were observed, with the effects being carried over into the fourth (non-fumigated) growing season. The results from this study suggest that the impact of ozone on biomass production may be cumulative and of a continuing nature.

Recent work by the Institute of Terrestrial Ecology at Edinburgh has shown that acid mists at realistic dose rates may reduce tree growth. The experiments involved exposure of 18 year old trees, of a single clone of Sitka spruce, to an equimolar mixture of H_2SO_4 and NH_4NO_3 at pH 2.5 twice weekly at a rate equivalent to 4 mm precipitation per week, between May and November 1990. A similar mist treatment was applied to 2 year old seedlings in nearby open-top chambers. The treatment produced no visible injury, but significantly decreased the relative height growth of 18 year old trees, and also reduced their relative stem area increment in all height classes, by up to 25%, compared with trees receiving no mist treatment. In contrast, in this and in earlier experiments, there was no significant detectable effect of acid mist on the growth of young trees. The reduction in stem volume of the older trees has been produced at a dose rate approaching that received by upland forests in several areas of Britain and points to the possibility of significant yield reduction in some

British forests. However, the control trees received no mist treatment and were never enclosed inside a misting chamber. Thus, although the extra water input from the mist was small and chamber enclosure was only intermittent, the possibility that the effect was not due to mist acidity cannot be excluded.

The Effects of Air Pollutants in Combination with Abiotic and Biotic Stresses

Over the last few years, evidence has accumulated to show that air pollutants can affect plant growth and development by altering important physiological processes which control the assimilation and utilization of carbon, water or nutrients (Darrall, 1989). Many of these processes are also influenced by natural stresses, in particular by water deficits, low temperatures and disease. There is now a body of experimental evidence to suggest that air pollutants interact with these stresses.

Water relations

The maintenance of favourable water relations is of fundamental importance to all terrestrial plants and several physiological responses have evolved to cope with water deficits. Two principal ways by which pollutants might alter the water relations of trees are:

(a) as a result of an imbalance in the allocation of carbon so that the growth of the shoot system is favoured over the root system, thus reducing the capacity of the roots to explore for further supplies of water and increasing the area of foliage available for transpiration.

(b) through effects on stomatal function and the control of water loss from the cuticle.

In addition to effects on stomatal processes, air pollutants may also influence cellular processes and cell water relations, although the two processes are intimately associated.

Effects of pollutants on carbon allocation

There is increasing evidence that air pollutants can cause a major reallocation of dry matter away from the roots to the shoots, often resulting in an increased shoot:root ratio. Although many of the experiments carried out to study changes in carbon allocation have involved relatively short-term fumigations (usually over a few weeks) of herbaceous species (UK TERG, 1988), a small number of experiments with trees have pointed to the same phenomenon.

Freer-Smith (1985) exposed silver birch (*Betula pendula*) during the first year of growth to concentrations of SO_2 and NO_2 ranging from 30 ppb to 90 ppb. He found that shoot:root ratio was significantly increased by SO_2, but there was no obvious effect of NO_2 on the partitioning of dry matter between shoot and root. This latter conclusion is similar to that reached from studies with herbaceous plants, viz. that NO_2 alone does not have a major influence on the allocation of assimilates to different plant parts. Increases in shoot:root ratios in trees have also frequently been noted with O_3 exposure (McLaughlin, 1985).

However, root biomass or shoot:root ratios may not necessarily be the most informative parameters for detailed studies of pollution-drought interactions, since only a fraction of the root system may be active in water absorption. In this context, measurements of the length or surface area of absorbing roots in relation to th area of foliage would be more relevant. Recent studies by Taylor and co-workers (Taylor *et al.*, 1989; Taylor and Davies, 1990) used this type of approach in experiments with beech and Sitka spruce seedlings grown in unfiltered air in open-top chambers at the Forestry Commission's Headley site in southern England. During the 1988 growing season ambient concentrations of O_3 exceeded 60 ppb on 33 occasions and reached 85 ppb on at least 5 occasions. Concentrations of SO_2 and NO_2 rarely exceeded 25 ppb. Exposure to the ambient pollution resulted in a slight (but non-significant) reduction in root biomass for beech and a significant decrease in the shoot:root ratio of Sitka

spruce. This is contrary to most previous findings reviewed earlier (Mansfield, 1988). Both species, however, consistently showed an increase in root length in the unfiltered treatment and thinner roots were produced by the beech, particularly in the lower soil layers. At a later stage of the experiment, the imposition of a drought by withholding water from half of the beech trees inhibited the length of root produced, irrespective of whether the trees had been grown in filtered or unfiltered air.

There are contrasting theories as to the cause of changes in carbon allocation. Increases in shoot:root ratios may either be due to effects on the processes responsible for carbohydrate translocation, or as a consequence of preferential stimulation of above-ground growth. Lorenc-Plucinska (1986) investigated the effects of SO_2 on translocation in SO_2-sensitive and SO_2-tolerant seedlings of Scots pine. Short-term (5 h) exposure to high concentrations of SO_2 (250–750 ppb) inhibited the transfer of newly assimilated ^{14}C from the needles to the roots irrespective of the concentration of SO_2. Compared to photosynthesis, the translocation process appeared to be more sensitive to SO_2, both in terms of immediate effects and in the continued response after fumigation had ceased. Although a reduction in the quantities of sugars and amino acids exported from the needles to the roots was noted, no qualitative changes in the types of substances translocated to the roots were observed. Changes in translocation may also occur as a consequence of increased assimilate demand by the foliage. Since both SO_2 and O_3 are known to inhibit net photosynthesis (Darrall, 1989), this may explain why they both can cause a redistribution of photosynthates to developing leaves at the expense of the roots.

A preferential stimulation of above-ground parts and changes in shoot physiology associated with increased nitrogen inputs to tree canopies has been suggested as a possible inciting factor responsible for the forest decline in parts of central Europe and N. America. Greenhouse experiments with red spruce (*Picea rubens*) seedlings exposed to simulated acid rain (pH 4.1) or acid mist (pH 3.6) have shown that the drought sensitivity of this species was increased (Norby *et al.*, 1986). The increase in sensitivity was not attributed to any changes in physiological resistance to drought, but to the stimulation of transpirational water loss from increased foliar biomass. Norby and co-workers (Norby *et al.*, 1989) have also shown that ambient concentrations of oxides of nitrogen (NO_x) induced the production of the enzyme nitrate reductase in red spruce needles. The needles were, therefore, able to utilize the atmospheric nitrogen directly and this may provide a mechanism for foliar fertilization and a possible basis for changes in shoot to root ratios.

Effects of air pollution on stomatal function and water loss from the cuticle

Plant leaves are protected against excessive water loss to the atmosphere by an epidermis, the surface of which is covered by a cuticular layer composed of waxes. Water loss from individual leaves is controlled by the opening or closure of the stomata, which in conifers are characterized by an extension of the cuticle in the form of a porous plug of wax tubules which cover the deeply sunken stomata.

In forest stands, the control of water loss and gas exchange is also determined by tree canopy structure and by boundary layer resistance (Jarvis and McNaughton, 1985; Jarvis, 1985). In the UK, no experiments have been carried out to investigate the effects of air pollutants on tree water relations at the stand level. In Germany, Schulze *et al.*, (1987) found that differences in the water relations between a declining stand and a healthy stand of Norway spruce were not related to canopy transpiration but to restricted root growth and distribution, so that despite a higher total rainfall the declining stand was more prone to drought.

Stomatal function

Because of the problems associated with exposure of mature trees in the field, most measurements of the effects of air pollutants on stomatal function

have been restricted to laboratory studies with plants grown under controlled conditions. The effects on stomatal function have been well documented (Darrall, 1989). Fumigation with high concentrations (>1000 ppb) of SO_2 or O_3 resulted in stomatal closure, concentrations in the range 100–1000 ppb resulted in either closure or no response and lower concentrations often induced stomatal opening. This suggests that at these pollutant levels, plants would be less able to control their water loss and this would affect their ability to withstand periods of water shortage.

It is often the case that the meteorological conditions favourable to the formation of ozone (UK PORG, 1987) can also lead to the development of soil moisture deficits. The concurrence of ozone and drought conditions and the observations that the decline in forest health in parts of central Europe appeared to be accelerated during drought years. This has led to the hypothesis that ozone may be a factor which contributes to tree damage by causing an increase in transpiration (Prinz *et al.*, 1987). In the UK and elsewhere, the majority of experiments to investigate the effects of O_3 on transpiration and water relations of trees have been conducted on well-watered seedlings. Surprisingly there is only very limited information concerning the interaction between ozone and drought applied simultaneously. Where the latter type of study has been conducted, water stress has usually been found to reduce the direct impact of a pollutant by causing stomatal closure and thus decreasing the uptake of the pollutant (Freer-Smith and Dobson, 1989). This suggests that drought events of a limited duration may offer some protection from pollutant damage as a result of reduced stomatal conductance. This is in agreement with a recent study by Dobson *et al.*, (1990) where Norway and Sitka spruce were exposed to episodes of O_3 (80–100 ppb for 2–3 hours) concurrent with the imposition of a moderate drought. Although undetermined, it is likely that under conditions of severe water deficit the degree of protection from pollutants afforded by stomatal closure would be offset by a reduction in CO_2 uptake and photosynthesis.

Significant interactions between O_3 and soil water stress were found on the growth and gas exchange of beech seedlings by Davidson *et al.* (1992). Well-watered seedlings showed a decrease in growth with increasing O_3 dose over a range of seasonal exposures characteristic of different years in southern Britain (20–48 ppb 8 h season mean). In contrast, seedlings experiencing a mild soil water deficit showed an increase in growth with increasing O_3 dose. Gas exchange measurements also showed that, on some occasions, stomatal conductance increased with increasing O_3 dose in water-stressed, but not well watered, seedlings.

In the case of well-watered seedlings it has been demonstrated by Freer-Smith and Dobson (1989) that fumigation of Norway spruce and Sitka spruce with O_3 at a concentration of 80 ppb could result in an increase in transpiration of about 25% compared to control trees in filtered air. Similar results were also obtained by Keller and Hasler (1988) for Norway spruce, by Bucher *et al.* (1988) for fir (*Abies alba*) and by Skarby *et al.* (1987) for Scots pine. In all these experiments transpiration rate and stomatal conductance were measured immediately following or during a period of ozone fumigation. However following experiments (Barnes *et al.*, 1990; Eamus *et al.*, 1990) in which Norway spruce was exposed to O_3 for 3 consecutive summers (8 hr mean 80 ppb), daily transpiration rates and daytime stomatal conductance of Norway spruce shoots were measured approximately 7 months after the end of the third year of pollutant exposure. It was found that transpiration from 1-year-old needles was increased by 28% and from current year foliage by 16% compared to shoots from charcoal filtered air, while stomatal conductance to water vapour of 1-year old needles was increased. Similar results have also been recorded by others, for example Reich and Lassoie (1984) for hybrid poplar (*Populus deltoides* x *trichocarpa*). In this study a 2 month exposure to ozone (mean concentration 125 ppb) resulted in a lowered leaf water use efficiency (defined as the amount of water consumed per unit of carbon gained in photosynthesis) and it was concluded that this

was caused by an impairment of stomatal control. Stomatal closure was poor and the response sluggish in excised O_3-exposed leaves which also wilted much sooner than comparable control leaves. A similar decline in the responsiveness of stomata following excision of the shoot, and hence the imposition of a severe and rapid water deficit to the needles, was also observed by Barnes *et al.* (1990) for O_3 fumigated Norway spruce. In this case water loss was also found to be related to the age of the needles. Whereas 2-year-old needles lost water more rapidly than 1-year-old, these in turn dried out faster than current year foliage.

The observations that the effects of O_3 on needle water loss may be long-lasting suggest that this pollutant may predispose conifers either to subsequent drought events, perhaps even in the following year, or to winter desiccation. Neither type of stress would be likely to lead to the death of mature trees but damage could be of a more localised nature, for example an accelerated aging of the older needles and premature needle fall.

Under controlled conditions other gaseous pollutants have also been found to promote the premature fall of leaves (Freer-Smith, 1985) and to increase water loss from seedlings of deciduous trees (Neighbour *et al.*, 1988). In this latter experiment, birch species (*Betula pendula* and *B. pubescens*) were exposed to a range of concentrations of SO_2 and NO_2 in combination. After 40 days exposure to the pollutant treatments, the ability of the leaves to control water loss under severe stress was tested by detaching the leaves and weighing them at intervals over several days. The leaves from the control plants and those exposed to 20 ppb SO_2 + 20 ppb NO_2 lost about 15% to 20% of their water content by the end of the first day, whereas the leaves from plants exposed to either 40 ppb SO_2 + 40 ppb NO_2 or 60 ppb SO_2 + 60 ppb NO_2 lost nearly 40% of their water content within 2 to 3 hours of being excised (Fig. 5.3). Exposure to the two higher pollutant treatments also resulted in an increase in transpiration throughout the day and night and the premature senescence and abscission of leaves.

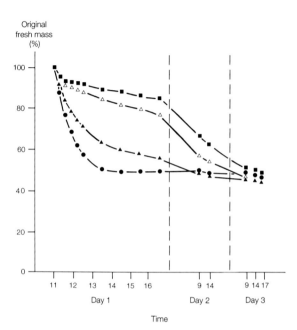

5.3 Change in weight over time of leaves of *Betula pubescens* exposed for the previous 30 days to ambient air and to mixtures of SO₂ and NO₂ at different concentrations.
((■) – ambient air; (△) – 20 ppb SO₂ + 20 ppb NO₂; (▲) – 40 ppb SO₂ + 40 ppb NO₂; (●) – 60 ppb SO₂ + 60 ppb NO₂). Individual values are means of 9 replicates. (From Neighbour *et al.*, 1988.)

Few detailed experiments have been conducted to investigate the direct effects of acid mist or rain on those aspects of tree physiology concerned with the control of water loss and drought susceptibility. Neufeld *et al.* (1985) found that in three of the four deciduous tree species native to N. America which they investigated, no change in stomatal conductance was observed following exposure to acid rain of pH 2.0, but the conductance of a fourth was reduced. Similarly Taylor *et al.* (1986) found that transpiration of red spruce (*Picea rubens*) was unaffected by acidic rain or mist of pH 3.6. In contrast, a recent study by Eamus and Fowler (1990) showed that exposure of red spruce seedlings to acid mist of pH 2.5 resulted in increased stomatal opening at low light fluxes and similar effects have also been observed by others (Fluckiger *et al.*, 1988; Barnes *et al.*, 1990).

Cuticular water loss

Waxes, either in the cuticle, or present as plugs in the stomatal antechambers of conifers, represent the main barrier to cuticular water loss. In recent years it has been speculated that air pollutants may alter the structure and/or composition of these waxes (reviewed by Turunen and Huttunen, 1990). Studies on enzymatically isolated cuticles from both evergreen and deciduous leaves have, however, shown that at realistic concentrations only prolonged exposure to NO₂ appeared to increase the permeability to water vapour (by a factor of 2–5) (Lendzian *et al.*, 1986; Kersteins and Lendzian, 1989). For the majority of trees the magnitude of cuticular conductance is of the order of 10–50 times lower than stomatal conductance and therefore represents an insignificant proportion of the total leaf conductance when the stomata are open or even partially closed. Because incomplete stomatal closure will also contribute to water loss when stomata are present on a leaf surface, it is very difficult to measure cuticular water loss. Hence, although there have been several reports which suggested that exposure to air pollutants could increase 'cuticular transpiration' in conifers (Fowler *et al.*, 1980; Mengel *et al.*, 1989; Barnes *et al.*, 1988), a more likely explanation is impaired stomatal regulation caused by pollution-induced structural degradation or premature ageing of the epicuticular waxes (Grill *et al.*, 1987; Riding and Percy, 1985; Barnes *et al.*, 1990).

Effects of air pollution on cell water relations

While the cuticle and stomata are the initial receptors of air pollutants, once through the stomata gaseous pollutants can diffuse into the substomatal intercellular spaces. Here acid-forming pollutants such as SO₂ and NO₂ can dissolve at the moist surfaces of the mesophyll tissues to produce a variety of reactive molecules (sulphite, bisulphite or nitrite ions, or in the case of O₃, free radicals). Reactions both within the cell and at the cell wall might result in potentially important changes to parameters

44

which are associated with cell water relations and hence the ability of the tree to resist or reduce the impact of drought episodes. Although only a limited amount of information is available at present, much of the evidence from conifers points to a reduction in maximum cell turgor following exposure to acid mists of pH 2.5 (Eamus *et al.*, 1989) or pH 3.6 (Barnes *et al.*, 1990) or to NO$_2$ (Freer-Smith and Mansfield, 1987). However, exposure of Norway spruce to O$_3$ and/or acid mist has also been shown to cause an increase in the elasticity of cell walls (Barnes *et al.*, 1990) which suggests that in some cases changes in the turgor relations of fumigated tissues may compensate partly for physiological changes in stomatal function.

Frost damage and winter injury

Damage to trees caused by low temperatures is an important factor in UK forestry *(see Chapter 4)*, and can cause widespread damage to a range of tree species. Such damage has complex causes, and there are a number of potential ways in which pollutants may alter the sensitivity of trees to low temperature stress (Davison *et al.*, 1988). The three most important considerations are

1. In mid-winter, dessication damage to evergreens is at least as important as direct frost damage. Major air pollutants are known to influence stomatal functioning, and accelerate cuticular erosion *(see above)* and thus may reduce the ability of trees to withstand winter dessication.

2. Membrane damage is thought to be a primary cause of frost injury. Air pollutants are known to alter the fatty acid and protein composition of cell membranes and, in particular, to influence the activity of enzymes responsible for trans-membrane solute transport.

3. Much direct frost injury occurs not in mid-winter, when the plants are effectively hardened, but in the autumn, before trees are fully hardened and more particularly in the spring, when a warm period promoting de-hardening

or budburst is followed by a sudden frost. Pollutant exposure, by altering the temporal changes in physiology over the autumn and spring periods could significantly alter the extent of frost damage at these times.

There is a considerable body of evidence to show that exposure to SO$_2$, even at relatively low concentrations, can influence the sensitivity of a range of species to cold stress. Furthermore plants are generally more sensitive to SO$_2$ and NO$_2$ when grown at low temperatures because of a reduced capacity to detoxify the pollutants (UK TERG, 1988). There is evidence from field studies around industrial areas in Finland and in high elevation forests in Czechoslovakia that interactions between air pollution and winter injury are an important factor influencing tree health in those countries.

More recent work in the UK has provided evidence that exposure to O$_3$ and to acid mists can also influence the sensitivity of trees to winter injury. Both types of pollution stress could be important because many plantation forests in the uplands are regularly enveloped in acidic mist, and high O$_3$ concentrations tend to persist through the night at high elevation sites where winter injury is more common.

The first indication of O$_3$/cold stress interactions originated from an observation by Brown *et al.* (1987). Some Norway spruce saplings which had been exposed to 150 ppb O$_3$ over the summer had been left in a ventilated 'solardome'; in the subsequent November, visible injury was observed on the older needles of three of the ten clones involved after a period when air temperature dropped to −7°C. However, there was little evidence of any effects on saplings exposed to 50 or 100 ppb O$_3$ over the summer. This observation led to more controlled experiments to investigate the phenomenon. Norway spruce saplings were exposed to 120 ppb O$_3$ for 2 months, followed by a 2 week hardening period, and then exposed to overnight temperatures between −6° and −18°C (Barnes and Davison, 1988). Visible injury developed on older needles of some of the eight clones tested, during the post-freezing recovery. On three of the eight clones, more visible injury

45

was found on the O₃ treated plants than on the control plants after they had been subjected to freezing temperatures, thus confirming that O₃ increases sensitivity to freezing injury. Lucas *et al.* (1988) conducted a similar three month summer exposure to O₃ (at concentrations of 70 ppb, 120 ppb and 170 ppb) before testing the frost hardiness of saplings of Sitka spruce. In this case, the saplings were allowed to harden naturally over the autumn period, in order to avoid possible artefacts introduced by the rapid hardening used by Barnes and Davison (1988). Frost hardiness was tested on excised shoots in early November and early December, at a range of temperatures. On the first occasion, there was significantly more visible injury on the O₃ treated shoots (Fig. 5.4) but this effect was not found on the second occasion, when the shoots were much more frost hardy. Thus there is evidence from this experiment that O₃ may cause a delay in the autumn hardening of the shoots, but not affect their hardiness in mid-winter. However, more recent data suggest that O₃ may also act by increasing night-time rates of transpiration (Lucas *pers. comm.*).

These studies indicate that O₃ can increase the sensitivity of tree seedlings to winter stress. However, all involve O₃ concentrations well in excess of those found in the UK, and all involve artificial freezing of excised shoots to assess frost tolerance. Senser and Payer (1989) reported that although O₃ at very high concentrations affected the frost resistance of Norway spruce and Scots pine seedlings, there was no evidence of any affect at lower, more realistic concentrations. Thus, although recent experiments have provided direct evidence that O₃ can influence the responses of young trees to winter stress, there is no evidence that O₃ actually has this effect under field conditions in the UK.

Studies on the exotic spruce species *Picea rubens,* have shown that exposure to acid mists at a pH of 3.5 or lower over the period July–December caused an increased frost sensitivity in the autumn compared with seedlings treated with pH 5 mist (Fowler *et al.*, 1989b). Further studies enabled the ions responsible for these changes in frost sensitivity to be identified by using mist solutions containing

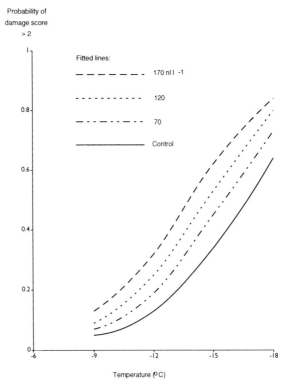

5.4 Relationship between freezing temperature and needle damage in Sitka spruce exposed to four O₃ concentrations. (From Lucas *et al.*, 1988.)

different combinations of hydrogen, ammonium, sulphate and nitrate ions. The results showed no effect of nitrate or hydrogen ion concentration. The delay, or reduction, in autumnal frost hardening was associated with increased concentrations of sulphate, and, to a lesser extent, ammonium ions (Cape *et al.*, 1991). Subsequent studies with Norway spruce seedlings have confirmed that similar responses are found with a spruce species of importance in the UK (Cape *et al.*, 1990). Treatment for one growing season with mists containing sulphate ions at a concentration of 1.6 mM resulted in shoot death at higher temperatures during the autumn (Fig. 5.5). However no effect on frost hardening was found when ammonium nitrate and nitric acid mists were applied. The same experiment also showed that over 200 hours exposure to 100 ppb O₃ during the growing season did not significantly affect frost hardening, although exposure to 140 ppb O₃ at the same frequency did.

46

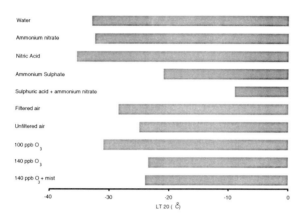

5.5 Freezing temperatures causing 20% shoot death in Norway spruce in response to acid mist or O_3 exposure. (LT_{20} – lethal temperature at which 20% shoot death occurs.) (From Cape *et al.*, 1990.)

Biotic stresses

Insect pests and plant pathogens can have major ecological or economic impacts on forestry in the UK *(see Chapter 4)*. These impacts may arise directly from the damage caused by the pest or pathogen, or indirectly, as a result of the ecological impact of pesticides applied to control them. Many, if not most, instances of forest decline involve attack by insect pests, or fungal pathogens. In some cases, these may be a primary cause of decline, while in others they are secondary factors which accelerate the decline of trees which have already been weakened by other factors. These factors may include air pollution; for example, the decline and death of *Pinus ponderosa* trees affected by very high ozone concentrations in parts of southern California is often accelerated by attacks by root rot fungi and bark beetles.

Recent field and experimental evidence has provided further evidence that air pollution can modify the incidence of such pests, although there is limited specific information relating to major UK tree species. For both insect pests and plant pathogens, it is important to recognise that air pollutants can act in a number of ways:– by a direct effect on the organisms themselves, by changes in leaf surface structure and chemistry, and by changes in leaf or phloem chemistry. In addition, when consider-

ing the population dynamics of insect pests in the field, the possibility of indirect effects, through changes in the activity of natural enemies and parasitoids cannot be discounted.

In general, direct adverse effects of air pollution on insect herbivores are found at relatively high concentrations (Reimer and Whittaker, 1989), and the effects of concern at current UK ambient concentrations are those mediated through changes in plant chemistry. The evidence for such effects takes a number of forms:–

1. Filtration studies in London (Dohmen *et al.*, 1984; Houlden *et al.*, 1991) have demonstrated that growth rates of aphid pests of agricultural crops on plants grown in unfiltered air are consistently higher than on plants grown in filtered air.

2. Observations alongside major roads have frequently shown high populations of herbivorous insects. For example, Port and Thompson (1980) reported outbreaks of defoliating lepidopteran larvae on beech and hawthorn planted alongside UK motorways, which they attributed to chemical changes induced by vehicle emissions.

3. Studies of grain aphid populations in an open-air fumigation of wheat and barley with SO_2 have demonstrated a positive correlation with aphid density over a range of mean seasonal SO_2 concentrations from 10ppb to 50 ppb. This was associated with an increased nitrogen content at high SO_2 levels (Aminu-Kano *et al.*, 1991).

4. Short-term fumigation experiments using 100 ppb SO_2 and NO_2 have clearly demonstrated that a range of important pests of agricultural crops show increased growth rates when on fumigated plants (Houlden *et al.*, 1990). Warrington (1987) showed that, in the case of pea aphids, growth rates increased linearly with SO_2 concentration applied for four days over the range 10–100 ppb SO_2, although above 150 ppb growth rate rapidly declined, probably because of a direct effect on the aphids.

5. Studies along natural pollution gradients have also shown significant effects of ambient air pollution. For example, Citrone (1989) found increased numbers of spittlebugs on thistles growing close to a coke works emitting SO_2.

All these studies provide a body of evidence to demonstrate that the levels of SO_2 and NO_2 found in the polluted areas of the United Kingdom can significantly increase the performance of aphids and other pests on herbaceous species. However, there is relatively little information relating to insect pests of trees in the UK, while the effects of pollutants other than SO_2 and NO_2 are poorly understood.

The clearest evidence available for an interaction between forest pests and air pollution is for effects on the green spruce aphid, *Elatobium abietinum*. This is an important pest of Sitka spruce in the UK, and which is intimately involved in the problems with Sitka spruce growth in South Wales *(cf. Chapter 6)*. Short-term fumigation of Sitka spruce seedlings with 100 ppb SO_2 for only 3 hours caused a 20% increase in *Elatobium* nymph growth rate (McNeill *et al.*, 1987) (Fig. 5.6). Increases above 10% occurred at fumigation times between 16 and 48 hours. Fumigations with 100 ppb NO_2 also caused a significant increase in *Elatobium* growth rate on Sitka spruce. In this case the response developed less rapidly, but after 1 week's fumigation, increases in growth rate of above 50% were found (McNeill and Whittaker, 1990).

The effects of longer-term exposure to SO_2 on both *Elatobium* performance and on aphid-induced effects on Sitka spruce growth were investigated by Warrington and Whittaker (1990). Fumigation with 30 ppb for two months over the spring, resulted in a three-fold increase in aphid density compared with the control treatment. There was evidence of a synergistic interaction of SO_2 and aphid infestation on some, but not all, growth parameters measured. In particular aphid infestation reduced root weight by 32% in the presence of SO_2, compared with 2% in the control treatment.

Studies of other important UK aphid pests of conifers (*Schizolachnus pineti* and *Cinara pini*

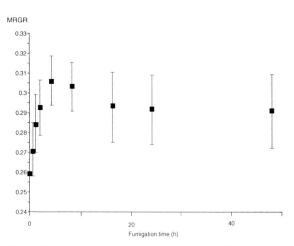

5.6 Three-day mean relative growth rate (MRGR) of the *Elatobium abietinum* nymphs on Sitka spruce prefumigated with 100 nl l⁻¹ SO_2. Vertical bars indicate 95% confidence intervals. (From McNeill and Whittaker, 1990.)

on Scots pine; *Cinara pilicornus* on Sitka spruce) have also shown increases in nymph growth rates in response to fumigation with 100 ppb NO_2 for a few days, (McNeill and Whittaker, 1990) but the effects of SO_2 on these pest/host plant systems have not yet been studied.

The experimental studies conducted to date on conifer aphids with O_3 suggest more complex pattern of response than that found for SO_2 and NO_2. There is evidence that the aphid response depends on the age of the plant material, the pattern of O_3 exposure and the temperature during fumigation. Both *Elatobium abietinum* on Sitka spruce and *Cinara pini* on Scots pine have shown no response to fumigation of host plants with 100 ppb O_3 for up to four days (McNeill and Whittaker, 1990). For *Schizolachnus pineti* on Scots pine, no response was found with episodic O_3 fumigation (for 8 hours each day) but aphid growth rate was reduced on plants fumigated with O_3 continuously. Experiments with *Aphis fabae* on *Vicia faba* have also shown differences in response between continuous and episodic O_3 fumigations (Brown and Bell, 1990). Responses of *Cinara pilicornis* grown on new shoots of Sitka spruce fumigated with 100 ppb O_3 were temperature

48

dependent, with positive responses being found at temperatures below 18°C, and negative responses at temperatures above 24°C. However, no significant effect of O₃ was found when older spruce shoots were used (Brown and Bell, 1990).

There are more limited data on the responses of insect herbivores on broadleaved species to air pollution. Studies at Lancaster University, involving five-day fumigations, found no effect of 70 ppb O_3 on *Eucallipterus liliae* on lime or 40 ppb SO_2 on *Drepanosiphum platanoides* on sycamore. However, exposure of birch to 105 ppb NO_2 increased the growth rate of *Euceraphis punctipennis* while exposure of beech to 40 ppb SO_2 increased the growth rate of *Phyllaphis fagi* earlier, but not later in the season (McNeill and Whittaker, 1990). Increased population growth rates of *Phyllaphis fagi* were also reported by Braun and Flückiger (1989) after growth of beech seedlings for two months in unfiltered air chambers, relative to those in filtered air, at a Swiss site where O_3 is the dominant pollutant. Interestingly, *Aphis fabae* population growth rates on *Phaseolus vulgaris* grown in the same chambers were decreased in unfiltered air suggesting again the variability of O_3 effects on plant/insect herbivore relations.

Braun and Flückiger (1989) also reported that acid mist at both pH 3.6 and pH 2.6 reduced the rate of population growth on *Phyllaphis fagi* on beech (Fig. 5.7). In contrast, Kidd and Thomas (1988) reported an increased population growth rate of *Schizolachnus pineti* on Scots pine sprayed with sulphurous acid mist.

In summary, there is evidence from these experiments that SO_2 and NO_2 at concentrations which are close to those found in the more polluted areas of the UK can increase the performance of some important insect pests. However, for O_3 the evidence suggests a more variable response, with both positive and negative effects on insect performance being reported, while for other pollutants there is little or no relevant data. Caution must be exercised about even these limited conclusions, since:–

a) Only a small range of insect herbivores have been studied, and these are mainly phloem

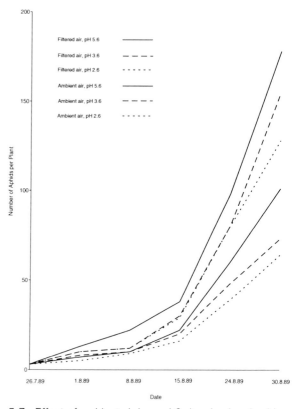

5.7 Effect of ambient air in rural Switzerland and acid mist, on the development of populations of *Phytaphis fagae* on beech (*Fagus sylvatica*). (From Braun and Flückiger, 1989.)

feeders. It should not be assumed that the responses of other insects to air pollutants will be the same.

b) Nothing is known of the responses of insect herbivores to pollutant mixtures.

c) Most of these experiments involve short-term fumigations under artificial conditions. The extent to which the long-term dynamics of insect populations is influenced by air pollution in the field is uncertain, and we do not know whether the effects would be of sufficient size to significantly influence forest vitality.

As in the case of insect pests, there are several mechanisms by which air pollutants may influence plant pathogens, the major ones being direct effects on the pathogen; effects on leaf surface structure and chemistry and effects on internal structure and chemistry (Dowding, 1988). Two important differences may also be noted: (i) direct effects on fungi occur at much lower pollutant concentrations than those on insect pests, and (ii) effects on the leaf surface are of much greater significance. It is well known that certain pathogens do not occur in areas affected by high concentrations of SO_2. For example, black spot (*Diplocarpon rosae*) of roses, and tar spot (*Rhytisma acerinum*) of sycamore were not found in areas of Britain experiencing SO_2 concentrations above 30 ppb SO_2 (Saunders, 1966; Greenhalgh and Bevan, 1976). These diseases are still much less prevalent in urban areas, although SO_2 concentrations have fallen substantially since the time these observations were made. There is some dispute as to the cause of this. For instance, Leith and Fowler (1987) claimed that the absence of tar spot from the centre of Edinburgh, where mean concentrations are now 15 ppb SO_2 was due to the clearance of fallen leaves in the city centre removing the major source of re-infection.

At relatively low SO_2 concentrations, the effects on disease development are complex. Variable effects may be found, for example, depending on the concentration used, the nature of the pathogenic organism, the timing of SO_2 exposure in the relation to the pathogen's life cycle, and the nature of the experimental system. Unfortunately, chamber systems, because of altered air flow rates and patterns and their removal or reduction of rain and mist, may greatly influence conditions on the leaf surface which may be critical to disease establishment. Thus the most reliable experimental data on the impact of air pollutants in field conditions come from open-air fumigation systems.

A study of disease development on a barley crop for three seasons in an open-air fumigation system with SO_2 shows the complex nature of the effects on disease development (Mansfield *et al.*, 1991). The effects of SO_2 on two major diseases differed completely. Powdery mildew (*Erysiphe graminis*) consistently increased in severity at elevated SO_2 levels, while leaf blotch (*Rhynchosporium secalis*) consistently decreased in severity at elevated SO_2 levels. Interestingly, this observation is in direct contradiction to the often-stated generalisation that SO_2 depresses the incidence of biotrophic organisms (such as powdery mildew) but not nectrophic organisms (such as leaf blotch). The effects were not linearly related to SO_2 concentration. Indeed powdery mildew tended to show a greater stimulation in the low SO_2 treatment (15 ppb) than in medium (30 ppb) or high (45 ppb) treatments. Further different patterns were observed on other pathogens, for example, diseases caused by *Fusarium*, *Puccinia*, *Septoria* and *Botrytis* species were unaffected by SO_2, while *Cladosporium* and *Alternaria* species responded in different ways in the different years.

Observations of surface micro-organisms on conifer needles have also been made in the Liphook open-air fumigation system. These have shown some evidence of reduced populations in the high SO_2 treatment. Laboratory studies have only shown effects of SO_2 on the organisms concerned at concentrations well above those used at Liphook, and reasons for this difference are not known.

There have been a number of studies in the U.S.A. demonstrating both positive and negative effects of O_3 on diseases of agricultural crops, and more recent European studies suggest that O_3 at realistic UK levels, could also affect such diseases. For example, Tiedemann *et al.* (1990) found that treatment with only 40 ppb O_3 for six days significantly inhibited the development of *Botrytis* on broad bean when the leaves were inoculated with the disease after fumigation but the treatment stimulated the development of *Septoria* on wheat.

Experimental evidence of an affect of air pollution altering the performance of pathogens of broad-leaved species was obtained by Flückiger *et al.* (1990). Tree seedlings were grown for one year in open-top chambers ventilated with ambient or filtered air at a site in Switzerland where O_3 is the dominant air pollutant. They were then artificially infected with pathogens, and the degree of leaf necrosis which developed was noted. In general, less necrosis developed on seedlings grown in ambient air. This was observed for leaf and twig infection with *Apiognomonia errabunda*, and for twig infection with *Nectria ditissima* on beech, and for leaf infection with *Apiognomonia veneta* on London plane. However, winter infection of twigs of London plane with *Apiognomonia veneta* resulted in more necrosis in trees grown in ambient air. Flückiger *et al.* (1990) also showed that fumigation with 50 ppb O_3 for 24 hours of conidia of *Apiognomonia veneta* grown on artificial media inhibited conidial germination.

There is thus evidence to suggest that current levels of SO_2 and O_3 in the UK may be high enough to influence fungal disease development. However, there has been virtually no experimental work on fungal diseases of major UK tree species, and the reported effects on other species are far too variable to permit any general conclusion even on the likely direction of the effects. The effects of other pollutants should also be considered. For example, acid mist and rain are known to affect both leaf surface structure and chemistry, which are known to be important influences on spore germination and disease establishment.

Multiple stress interactions

There is substantial evidence accumulating that air pollutants may modify the impact of other stress factors. Interactions between air pollutants and other stress factors have generally been considered separately. However, it is important to realise that in the field those stresses do not occur in isolation. For example, there is competition between microorganisms on the leaf surface, aphids act as major vectors for viral diseases while their honeydew can provide an important leaf surface substrate for microbial diseases. Holes punctured in the leaf by aphids may increase water loss and pollutant uptake rates (Warrington *et al.*, 1989) and plants under water stress may be more susceptible to attack by insect pests and fungal diseases. These are only a few of the potential interactions, and it is clear that any comprehensive evaluation of the indirect effects of air pollutants would need to go beyond simple two-factor experiments to consider the complex web of interactions within the forest ecosystem. Nevertheless, the major priority must

be to establish a stronger experimental basis on which to evaluate the impact of air pollutants at levels found in the UK on winter injury, plant water stress and disease severity in our forests.

Pollution-Induced Nutritional Problems

Soil acidification

There are several reasons why afforestation leads to soil acidification. Some of these were discussed in Chapter 2 and are related to enhanced deposition of acidifying pollutants. Others are associated with the accumulation of base cations in biomass and the development of acid organic horizons during forest growth.

Many data have now been published showing evidence of large increases in the acidity of forest soils in areas of Scandinavia, Germany, Austria, Switzerland and the UK over the last 40 years (Berden *et al.*, 1987; Billett *et al.*, 1988 & 1990). The measured increase in acidity is usually the result of the interactions of a number of factors. Part or at some sites the whole, of the increase results from the accumulation of chemical bases in the tree biomass, or from forest management, eg replacement of hardwoods with conifers leading to the development of an acidic forest litter. However, at other sites acidic deposition has certainly contributed to the acidification. Thus, for example, Hallbacken and Tamm (1986) concluded that changes in pH of the surface horizons of the soils they examined in Sweden were strongly influenced by forest growth but that increases in the acidity of the subsoil horizons were most probably due to the effects of acidic deposition.

Within the UK, Billett *et al.* (1988, 1990) examined soils from 15 sites in the Alltcailleach Forest, north east Scotland which had previously been sampled and analysed in 1949–50. Surface horizons from 80% of the soils showed a reduction in pH over the 40 years with a maximum reduction of 1.28 pH units, 73% of the subsoil horizons also showed a decline in pH, with a maximum reduction of 0.54 units. The authors concluded that "The key factors governing increases and decreases in soil Ph are changes in ground vegetation and tree canopy, although some effects of acid deposition cannot be ruled out".

Increased acidity *per se* is unlikely to affect tree growth directly, indeed several of the study sites referred to above show no recent adverse changes in tree health. However, acidification is usually linked to a reduction in available base cations. Thus, Billett *et al.* (1990) report that soil acidification in the Alltcailleach Forest was associated with a significant decrease in exchangeable calcium (Ca), magnesium (Mg), potassium (K) and sodium (Na).

Base cation deficiencies

There is a large amount of experimental evidence from both field and laboratory studies, showing increased leaching of base cations in response to irrigation with simulated acid rain (eg Abrahamansen and Stuanes, 1980; Lee and Weber, 1982; Morrison, 1983; Brown, 1987), or following exposure of soils to sulphur dioxide (Wookey and Ineson, 1991). The magnitude of the leaching, and of the impact on soil nutrient status varies with the chemistry of the inputs with soil properties. Major increases in leaching rates of base cations and significant impacts on base saturation are seen only when the leaching solutions have a pH < 3.0. Such low pH values are rarely reported for bulk precipitation in the UK. However, enhanced leaching has also been found following addition of neutral solutions containing large concentrations of mobile acid anions to acidic soils (Hultberg *et al.*, 1990; Matzner *et al.*, 1983). Large inputs of ammonia (NH_3) can also result in greatly enhanced leaching and rapid soil acidification, if nitrification produces nitrate (NO_3^-) which is not utilized by the soil–plant system.

The results from the experimental studies, combined with existing models of soil leaching mechanisms and weathering processes, can be used to predict the response of soils to acidic inputs (Wang and Coots, 1981; Catt, 1985; Wilson, 1989). In soils containing free carbonates, or with a large cation exchange capacity and high base saturation, the enhanced leaching does not have a significant impact on soil base status in the short to medium term. Soils of this type occur over large areas of southern and eastern England (Fig. 5.8). At the other extreme, in very acid mineral soils with low base saturation ($< 10\%$), acidic inputs mainly result in increased mobilization of aluminium (Al) but with some additional loss of other cations. Soils of this type occur over large areas of the British uplands (Fig. 5.8). The possible impacts of large concentrations of Al in soil solution on tree growth are discussed later. The most sensitive soils to cation depletion in the short to medium term are those with small cation exchange capacities and moderate base saturation (10–25%). In the UK, soils of this type mainly occur on coarse textured soil materials. Their distribution is shown in Fig. 5.8. However, the possible impact of enhanced leaching of base cations must be evaluated alongside recent and current atmospheric deposition. The assessment must include inputs of protons, (H^+), NO_3^-, sulphate (SO_4^{2-}), NH_3 and base cations. Many areas in western Britain with soils that are vulnerable to base depletion receive an input of base cations from marine sources that is equal to or greater than the annual requirements for growth.

The Type 1 spruce decline observed in central Europe (Rehfuss, 1987) has been associated with low foliar concentrations of Ca, zinc (Zn) and, in particular Mg. The cause of the Mg deficiency and whether it is a primary cause of the decline or a secondary effect has been vigorously debated, but there now seems to be a consensus that it is a major factor in the decline. The reasons for the reduction in the Mg status of the trees is still uncertain but pollution seems to be implicated (Roberts *et al.*, 1989). Increased inputs of acidic anions (SO_4^{2-} and NO_3^-) in acidic deposition may have led to increased leaching of Mg from the forest soils thus reducing the supply available for uptake. In addition, increased atmospheric inputs of nitrogen may have had a fertilizer effect, leading to increased growth which has outstripped the ability of the acidic soils to supply further Mg. Roberts *et al.* (1989) have pointed out that a similar widespread Mg deficiency is unlikely in the UK because atmospheric inputs of Mg, primarily from marine sources, are much greater than in central Europe.

Imbalance in cation uptake

Trees take up the majority of their base cation requirements from soil solutions, some passively with the transpiration stream and some actively. The active root mechanism allows some selective uptake of ions but the broad composition of the solution taken up reflects the chemistry of the soil solution. This latter is controlled by atmospheric inputs, modified by reactions with the vegetation canopy, and solution-soil solid phase reactions, which reflect the inputs and the properties of the soil phase. In areas with high pollutant levels of NH_3, the dominant cation in atmospheric inputs and soil solution is ammonium (NH_4^+). Root uptake of other cations, such as Mg, Ca and K is effectively suppressed and deficiencies can develop. Very large inputs of NH_3 (> 100 kg ha^{-1} yr^{-1}), high NH_4^+ : Ca and NH_4^+ : Mg in solution, and linked deficiencies have been reported in the Netherlands. (Draaijers *et al.*, 1989; can Breeman *et al.*, 1989). Experimental studies have also demonstrated cation deficiencies linked to an imbalance in uptake associated with ammonium dominance (van Dijk *et al.*, 1989 & 90). Similar large inputs of ammonia and induced cation deficiencies have not been demonstrated in the UK to date. Data on spatial variations in atmospheric concentrations of ammonia in the UK are poor at present but new data should soon become available. Large atmospheric concentrations of ammonia are associated with areas of intensive stock rearing and so the areas at risk from this type of induced deficiency could readily be identified.

A number of forest sites in the uplands of the UK have been shown to have an unexpectedly large output of nitrate in soil drainage and stream-waters (Stevens and Hornung, 1988; Stevens *et al.*, 1989). In Central Europe, increased nitrate leaching has been shown to follow forest decline and may imply that a nutritional imbalance is probable at the UK sites (Hauhs, 1989).

Reduction in mineralization

At an early stage of research into the impacts of acid deposition, it was suggested that acidic inputs and an increased soil acidification would suppress mineralization of organic nitrogen as the micro-organisms involved in the process were intolerant of very acid conditions. A large number of experimental studies have been done to test the hypothesis. These show that significant reductions in the rates of nitrogen mineralization are found only following irrigation with simulated acid rain at pH < 3.0, (Stroo and Alexander, 1986; Bewley and Stotzky, 1983; Like and Klein, 1985). Such acidic rainfall is rare in the UK and it seems unlikely that this mechanism would lead to a large enough reduction in nitrogen availability to produce an effect on tree health. Any such reduction in production of inorganic nitrogen in soils must also be compared with increased atmospheric inputs of inorganic nitrogen. However, data presented by Billett *et al.* (1990) suggests a build up of organic matter in the Alltcailleach Forest, north east Scotland over the last 40 years which cannot be explained as a normal result of forest growth. This accumulation of surface organic material is associated with a widening carbon/nitrogen ratio. These data suggest that there has been a reduction in decomposition rates, and require further investigation to identify the possible causes.

Aluminium toxicity

One of the earliest hypotheses to explain forest decline in Germany was that large soil solution concentrations of Al, induced by pollutant inputs to acid soils, were causing root damage leading to reduced nutrient uptake and forest decline (Ulrich *et al.*, 1980). Field observations in some German forests showing tree damage, suggested extensive death of fine roots and the symptoms were said to be similar to those of roots killed by Al in growth experiments (Matzner and Ulrich, 1985; Ulrich and Matzner, 1986). Soil solution ratios of Al : Ca of > 1 would, it was suggested, lead to root damage. Experimental studies have demonstrated that additions of simulated acid rain to acidic soils lead to enhanced mobilization of Al into soil solutions (eg Adams *et al.*, 1990). Experimental growth studies have also shown impacts on growth, fine root branching and fine root biomass in a number of tree species at relatively small concentrations of Al, (eg Paganelli *et al.*, 1987; Sucoff *et al.*, 1990; Kelly *et al.*, 1990). However, there is a considerable variation between tree species in their sensitivity to Al; a recent review by Raynal *et al.* (1990) suggests that broadleaved species are more sensitive than conifers, and spruce more sensitive than pine (Table 5.1). Ryan *et al.* (1986) showed that Douglas fir, Sitka spruce and western hemlock, which are amongst the more widely grown exotic conifers in the UK, are very tolerant of aluminium in solution. The Al: Ca ratio at which root damage has been reported also varies considerably; thus damage was found in red oak at Al : Ca ratio > 0.25 (Kelly *et al.*, 1990) but at > 1 in Norway spruce (Rost-Siebert, 1983). The mode of impact of the Al is still debated. Enhanced concentrations of Al in growth media are usually associated with reduced concentrations of Mg, Ca and often phosphorus (P) in foliage (Hecht-Buchholz *et al.*, 1987; Ryan *et al.*, 1986) and root tissue. Aluminium ions very effectively block Ca and Mg uptake (Stienen and Bauch, 1988) and Al has been shown to reduce calcium binding to root cortical cells of Norway spruce (Schroder *et al.*, 1988). It is clear that not all forms of Al in soil solution are toxic to roots. In particular Al-organic complexes seem to have little, if any effect on root structure and vitality. Although many upland soils in the UK have high soil solution concentrations of Al, much of this Al in near-surface horizons is present as organic complexes. However, the high levels of Al in some soils may block uptake of cations.

Impacts on mycorrhizae

Mycorrhizae, the symbiotic association of plant roots and fungi, are an important means by which trees acquire nutrients from inorganic and organic sources in soil. They are also involved in plant water relations and as defence against root pathogens. Acidifying pollutants can affect the abundance and nature of mycorrhizal associations on tree roots directly by effects on soil chemistry and indirectly, by affecting the distribution of assimilates to the roots. The effect of pollutants on mycorrhizae has recently been reviewed by Jansen and Dighton (1990). Reduction in soil pH and consequent increased mobility of toxic metal ions (particularly aluminium), resulting from acidic pollution (SO_2, acid rain and NH_4^+), have been shown to reduce tree root growth and mycorrhizal health in Germany (von Becker, 1982). Cortical and xylem cells of damaged roots were found to contain a Ca:Cl ratio of less than 1, compared to a value of greater than 1 for healthy roots (Bauch, 1982). In addition to reduced mycorrhizal development, soil acidification has been shown in UK experiments to lead to a change in the species composition of mycorrhizae on roots (Dighton and Skeffington, 1987). Although we have little data on the comparative efficiencies of different mycorrhizal species in nutrient uptake and enzymatic capacities, this shift in population could have implications for tree nutrition.

Theoretically, indirect effects of pollutants, reducing the amount of assimilate available to support adequate mycorrhizal infection and hence altering tree nutrition, are important. From the current, limited data available, the results are inconclusive (Keane and Manning, 1987).

It has been suggested that toxic concentrations of ammonium in soil solution lead to damage to mycorrhizae of trees and eventually to dieback of fine roots (Mohr, 1986). Reduced growth of pine seedlings and a reduction in the rate of mycorrhizal infection in spruce have been demonstrated in greenhouse studies. However, the concentrations of ammonium used in these and similar

Table 5.1: Thresholds for statistically significant reductions in root growth by Aluminium for pine (*Pinus*) and spruce (*Picea*) species as determined in solution, sand and soil culture experiments with seedlings. Variables measured were root biomass production unless otherwise indicated. A greater than symbol (>) indicates no significant reduction was observed at the highest concentration of aluminium tested.

(from Raynal *et al.*)

Study	Species	Al Threshold mM l^{-1}	Medium
Pine species			
Hutchinson *et al.* (1986)	*Pinus banksiana*	1.48	Sand
Eldhuset *et al.* (1987)	*Pinus strobus*	2.96	Sand
Hymphreys & Truman (1964)	*P. sylvestris*	3.0–5.0	Solution
McCormick & Steiner (1978)	*Pinus radiata*	> 0.74	Solution
	Pinus rigida	4.44	Solution
	*P. sylvestris**	4.44	Solution
Williams (1982)	*P. virginiana**	4.44	Solution
	*Pinus clausa**	1.22	Solution
	*Pinus taeda**	1.22	Solution
Thornton *et al.* (1987) Expt. I unpublished data	*Pinus taeda*	> 3.00	Solution
Thornton *et al.* (1987) Expt. II unpublished data	*Pinus taeda*	1.50	Solution
Paganelli *et al.* (1987)	*Pinus taeda*†	0.19	Sand
Joslin & Wolfe (1988), unpublished data	*Pinus taeda*	1.30	Soil
Spruce species			
Evers (1983)	*Picea abies*	> 1.50	Solution
Rost-Siebert (1983)	*Picea abies*	0.30$	Solution
Abrahamasen (1984)	*Picea abies*	0.74#	Solution
van Praag *et al.* (1985)	*Picea abies*	3.33	Sand
Makkon-Spiecker (1985)	*Picea abies*	2.96	Sand
Ingestad *et al.*	*Picea abies*	1.00	Solution
Hutchinson *et al.* (1986)	*Picea mariana*	0.37	Sand
	Picea glauca	0.37	Sand
Nosko *et al.* (1988)	*Picea glauca*	0.05	Sand
	Picea rubens	0.37	Sand
Schier (1985)	*Picea rubens*	3.70*	Solution
Thornton *et al.* (1987)	*Picea rubens*	0.25*	Solution
Ohno *et al.* (1988)	*Picea rubens*	> 0.54	Soil
Joslin & Wolfe (1988)	*Picea rubens*	0.25	Soil

* Growth variable was root elongation
† Growth variable was relative growth rate
$ Variable was root elongation; [Ca] = 0.20 mM
# Growth variable was total plant biomass; [Ca] = 0/09 mM

studies were much higher than any which have been reported from the field (Zottl, 1990).

There is little concrete evidence of pollutant-induced changes in soil chemistry adversely affecting tree health in the UK. However, comparison of soil conditions under beech trees with thin crowns with those under visible healthy trees growing at the same site has revealed some significant differences. Power and Ashmore (1991) made such comparisons at three acidic sites and three calcareous sites in southern Britain. At none of the six sites was there any significant difference in soil acidity beneath healthy and unhealthy trees. At the acidic sites, the soil Ca concentration was lower, and the Al/Ca ratio higher under unhealthy trees, but no significant differences in Ca concentration were found on calcareous sites under healthy or un-healthy trees. On both calcareous and acidic sites soil potassium concentrations were lower under un-healthy trees. On both types of site, unhealthy trees had a smaller proportion of live and mycorrhizal root tips. While these differences cannot necessarily be attributed to increased pollutant deposition, they do show that increased Al/Ca ratios are associated with poorer beech health on acidic soils.

Above and Below Ground Interactions

The major hypotheses involving pollutants suggested to explain forest decline phenomena have been crudely classified into 'bottom-up' hypotheses, in which the primary impact of pollutants is through changes in soil chemistry, and 'top-down' hypotheses, in which the primary impact is through effects of leaf tissue. It is important to realise, however, that these mechanisms may not operate in isolation. There is a limited body of evidence indicating that changes in the rhizosphere may influence the response of tree seedlings to air pollutants, and that physiological changes induced by exposure to air pollutants may influence the capacity of a tree to respond to soil acidification.

For example, air pollutants such as O_3 and SO_2 are known to reduce translocation of carbo-

hydrates to the root system and have, as described earlier, been shown to influence the size and structure of the root system and the degree of mycorrhizal infection. However, comparisons of pine seedlings with or without mycorrhizal roots have shown that the mycorrhizal symbiosis itself can provide protection against the adverse effects of O_3 and SO_2 on root growth and physiology (Carney et al., 1978; Mahoney et al., 1985). Thus the possibility exists of synergistic effects of air pollution and soil acidification on the root/mycorrhizal association.

Similar synergistic effects on tree nutrient balance may occur. Although leaf nutrient deficiencies of base cations are now thought to be associated with soil deficiencies rather than foliar leaching processes, there is some experimental evidence that O_3 can directly influence the rate of uptake of Ca and Mg, probably through effects on the transpiration stream (Jurat and Shaub, 1988). More importantly there is evidence from some, but not all, studies that plants grown in soils with reduced concentrations of base cations are more sensitive to adverse effects of air pollutants. For example, Klumpp and Guderian (1989) found that the adverse effects of 32 weeks exposure to air pollutant mixtures containing 75 ppb O_3 on photosynthesis of older needles of Norway spruce were much greater on seedlings grown in sand culture deficient in Ca and Mg. There is also a potential for effects of air pollutants in altering the rate of mineral cycling, for example, through changes in the rate of litter decomposition. Ineson and Wookey (1988) reported that several months exposure to only 14 ppb SO_2 reduced microbial activity in pine litter.

There are other potential interactions between above- and below-ground effects, for example through pollutant-induced changes in plant hormonal activity or on root aphids which have not yet been properly investigated. At the present time, no firm conclusions on many interactive effects can be drawn from the experimental data, but the possible role of interactions between pollutant effects above- and below-ground should be considered when seeking to explain field observations.

6

Recent Case Studies

Conclusions

- Direct damage by air pollution is known to occur close to known pollution sources, eg severe foliage damage of Norway spruce in northern England downwind of an intensive animal rearing unit.

- Instances also exist where pollution is probably the direct cause of damage to trees in areas remote from pollution sources, eg browning of foliage of trees and shrubs in eastern England in September 1989.

- Pollution may also be a contributory factor to damage on trees relatively close to industrial sources eg poor condition of Sitka spruce in South Wales and cannot be ruled out as a contributory factor in some situations where there is tree damage in areas remote from pollution sources, eg dieback of hedgerow ash.

- Instances exist where air pollution has been blamed for large scale damage but detailed studies have not confirmed this assertion, eg foliage browning of Sitka spruce in Britain in 1984.

Introduction

Air pollution has long been recognised as a cause of damage to trees in Britain. Evidence has been presented for the role of high levels of SO$_2$ in preventing the growth of Scots pine in parts of the Pennines during the 1950s and '60s (Farrar *et al.*, 1977) and there have been some striking records of severe local damage. A recent instance involved the death of fifteen hectares of Norway spruce

downwind of burning coal-mining spoil near the Forth-Clyde valley (Gregory *et al.*, 1989). Despite these examples, it has proved difficult to obtain widespread evidence of air pollution as a direct cause of damage. Support for this statement comes from the records of the nationwide disease diagnostic service operated by the Forestry Commission for both woodland and non-woodland trees. During the last 30 years, only 16 cases of definite or probable air pollution injury were diagnosed out of a total of 10,500 cases to which a definite or probable cause of damage was ascribed. This represents 0.15% of the total. Its role as an indirect cause is much more difficult to determine as it can only be elucidated by means of detailed study. Even then, the evidence is often circumstantial.

This section of the report is concerned with case histories of recent investigations of tree damage. In some cases the direct effects of air pollution were involved. In others, the role of air pollution was not immediately obvious, but, for one reason or another, it was given serious consideration, either as the principal cause of the damage or as an important contributory factor.

Ammonia Pollution in North Yorkshire

In the Netherlands, ammonia emissions from intensive animal production units have been associated with forest damage, with both direct toxicity and soil acidification being implicated. Direct toxicity can be related to high concentrations of N in foliage and, in particular, leaf arginine content has been correlated with the extent of damage (van Dijk & Roeloffs, 1988). Soil acidification is associated with increased nitrification of deposited ammonium, a process which generates acidity and can lead to increased cation leaching (*see Chapter 5*).

A study was initiated in North Yorkshire in 1990, centred around an intensive animal rearing unit and aimed at quantifying N transfer from the farm to surrounding forests. Ammonia concentrations in the air were monitored, together with rainfall and throughfall chemistry.

The first six months of the study have revealed the importance of the farm in releasing ammonia and subsequent deposition to the forests downwind. Based on the data currently available, the deposition rates downwind of the farm are approximately 100 kg N ha^{-1} y^{-1}, whereas upwind they range from 30–50 kg N ha^{-1} y^{-1}. The Norway spruce stands associated with the highest deposition levels are showing marked signs of decline, with severe needle loss and necrosis (Plate 6.1). Stands with lower deposition rates do not show these symptoms.

At this stage the precise cause of damage is not clear. There is a strong relationship between SO_4^{2-} and NH_4^+ enrichment beneath the canopies (Fig. 6.1) and both pollutants may be involved. The data provide strong evidence to support the hypothesis that co-deposition of NH_3 and SO_2 is occurring (McLeod *et al.*, 1990).

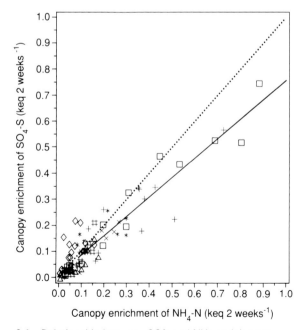

6.1 Relationship between SO_4^{2-} and NH_4^+ enrichment beneath conifer canopies around an intensive animal rearing unit in North Yorkshire. The symbols represent the different canopies, with the dotted line showing 1:1 equivalent and the solid line the actual regression.

Foliar Damage to Trees and Shrubs in Eastern England, September 1989

During September 1989, widespread foliar damage to a range of deciduous trees along the east coast of England from Scarborough to Kent was reported (Fowler and Pitcairn, 1991). The damage affected many species, but hawthorn (*Crataegus monogyna*) and blackthorn (*Prunus spinosa*) were most severely affected (Plate 6.2). The damage decreased with distance from the coast but was still clearly evident on east facing hedgerows in Lincolnshire and Suffolk 20 km from the coast.

The most severe foliar injury was observed close to the Lincolnshire coast, but widespread reports of damage were received at other sites along the east coastal regions of Yorkshire, Lincolnshire, Norfolk, Suffolk and Kent. From these, and a detailed survey in Lincolnshire, the area in which damaged vegetation was present has been estimated at 7600 sq km (Fig 6.2). Using the ITE land use data base and classification system it is estimated that the total linear extent of damaged hedgerow in these areas was 27000 km.

The cause of the foliar lesions is a complex issue. The period immediately preceeding the incident was warm and sunny, with a pronounced photochemical episode in south east England. Ozone concentrations reached phytotoxic levels of 100 ppb at Halvergate, a site close to the Norfolk coast. These large ozone concentrations, while potentially damaging to plants, are not consistent with the nature of the observed foliar injury and may be discounted as a likely cause. The damage was first observed at Halvergate close to the east coast of Norfolk where a large field experiment on air pollution deposition was in progress. Three days before the damage was noted there had been a day of moderate easterly winds and very low cloud and mist. On the morning following the 'mist', corrosive damage and a thin film of black particulate material was observed on the windward sides of polished alloy tubing associated with the experimental equipment. The atmospheric trajectories reaching the Lincolnshire and Norfolk coast on the day of the 'mist' showed that the air had passed over areas of the Netherlands and industrial regions of West Germany in the preceeding days. There were widespread reports of 'air pollution' damage to horticultural crops in the Netherlands during this period (van den Eerden *et al.*, 1990), and the air was shown to be very contaminated by the major pollutants SO_2, NO_2, O_3 and particulate material.

The presence of a toxic waste incinerator ship in the North Sea was also reported upwind of the Lincolnshire and Norfolk coastline during this period. The possibility that sea salt contributed to the observed foliar injury has also been examined.

Of the possible causes, sea-salt damage, while plausible from the form of the foliar lesions and geographical patterns of damage, was not consistent with the scale of leaf injury for the modest windspeeds during the few days prior to visible injury. The remaining possible causes were all in some way pollution-linked. The strong direction-dependent characteristics of damaged foliage were indicative of a droplet-borne agent and this contrasts strongly with conditions in the Netherlands where a combination of the gaseous pollutants, sulphur dioxide and ozone, appeared to be the cause of crop damage (van den Eerden *et al.*, 1990).

The concentrations of major ions SO_4^{2-}, NO_3^-, NH_4^+, H^+, Cl^-, and Na^+ in the mist would therefore appear to be the most likely cause. This mist was not collected by any of the national air quality monitoring networks and, as it was an unexpected event no specialized monitoring equipment was deployed during the field experiments noted earlier. However, large concentrations of SO_4^{2-} and Cl^- associated with the necrotic spots on leaves relative to undamaged leaves, supports this interpretation. The lack of elevated sodium ion concentrations in the damaged leaves supports the view that sea–salt damage was not the cause.

Table 6.1: Species on which damage was observed arranged into relative damage clases.

Severely damaged

Hawthorn	(*Crataegus monogyna*)
Blackthorn	(*Prunus spinosa*)

Moderately damaged

Sycamore	(*Acer pseudoplatanus*)
Birch	(*Betula pendula*)
Beech	(*Fagus sylvatica*)
Ash	(*Fraxinus excelsior*)
Willow	(*Salix spp*)
Horse Chestnut	(*Aesculus hippocastanum*)
Elder	(*Sambucus nigra*)
Lilac	(*Syringia vulgaris*)

Slight damage

Popular	(*Populus spp*)
Ornamental cherries	(*Prunus spp*)
Bramble	(*Rubus spp*)
Wild Rose	(*Rosa canina*)
Lawson Cypress	(*Chamaecyparis lawsoniana*)

No damage

Sea Buckthorn	(*Hippophae rhamnoides*)

Decline of Sitka Spruce in the South Wales Coalfield

Some of the forests planted on the South Wales coalfield in the 1960s are now in a state of decline. This is indicated by a reduction of leading shoot length from about 60 cm to as little as 8 cm, sometimes associated with bending of the shoot out of the vertical position. Other symptoms include thin crowns, dieback of shoots, various discolourations of the needles and the production of epicormic shoots. A multi-disciplinary effort was started in the early 1980s to investigate the causes of the decline. Aspects studied included soil hydrology and chemistry, atmospheric pollution, aphid outbreaks and tree growth and chemistry.

The affected forests are not far from heavy industry on the coast and from areas of much domestic coal burning. However, measurements of NO_2 showed mean annual levels typical of rural Wales (about 5 ppb). The mean annual concentration of SO_2 was about 10 ppb but monthly means of 20–40 ppb were recorded in the winter. The concentrations of both gases were too low to cause direct damage to Sitka spruce but may be sufficient to increase the susceptibility of trees to certain pests and diseases.

Detailed monitoring of crops on 20 sites encompassing a wide range of growth rates and tree health, indicated that growth rate was positively associated with foliar concentrations of certain elements, especially K and N. The strongest relationship found was between growth and low levels of K, although the latter were above what have normally been considered deficiency levels in Sitka spruce. The application of NPK fertilizer in a field experiment has shown a pronounced increase in growth and improved tree health. However, it is still too early to say whether the decline syndrome can be cured permanently by improved nutrition.

Stem analysis in declining crops showed that notable reductions of growth in certain years were associated with known outbreaks of the green spruce aphid in South Wales. All of the measured trees on the selected site were affected in the outbreak years but some recovered while others went into decline. The aphid is known to benefit from SO_2 pollution *(Chapter 5)*, although it is not certain whether the concentrations recorded in South Wales were sufficient to have an effect.

The work has led to the tentative conclusion that the decline of Sitka spruce on the South Wales coalfield has been caused by heavy defoliation (mainly caused by aphids although needle fungi may be involved) of trees of low nutritional status. The role of the recorded pollution levels on the defoliating organisms and on tree nutrition is unknown. A review of all aspects of this problem is being prepared for publication (see Coutts, 1993).

Dieback of Hedgerow/Ash

The phenomenon of dieback in hedgerow ash (Plate 6.3) has been known for decades but is poorly understood. Thus Pawsey (1983) reviewed a number of factors which might be involved in the

development of the condition. These included agricultural practices, atmospheric pollution, poor site and climatic conditions, pathogenic infections and insect infestations. Pawsey (1983) suggested that a complex of interacting factors was probably responsible for the initial development of dieback in trees, with other more secondary agencies influencing its further development. A recent survey (Hull and Gibbs, 1991) has shown that the condition is to be found mainly in the east of the UK, being most severe in the East Midlands where it was first noticed some 40 years ago (Fig. 6.3). Strong evidence has been found to provide confirmation for earlier suggestions of a link between damage and intensive arable farming, and also with factors such as the proximity of ditches and roads. In general, little evidence was found of a relationship with levels of atmospheric sulphur, nitrogen or ammonia. However with trees growing adjacent to arable land, there was a correlation between the severity of the condition and levels of atmospheric pollutants. Lack of data on ozone episodes precluded an examination of this pollutant.

Beech Decline

Work on assessment of the health of beech through the study of crown conditions is covered in Chapter 3. Reference should be made here, however, to a detailed analysis of shoot growth conducted by Lonsdale *et al.* (1989), which is seen as a sensitive measure of the tree's response to environmental influences. In their investigation Lonsdale and co-workers recorded annual shoot growth on samples of trees at 15 sites along a transect from south-west to east England and found that trees on seven of the sites had entered a period of declining growth at various times between 1971 and 1979 (Fig. 6.4). Although prolonged droughts had clearly had a major adverse influence on shoot growth, it was found that even when the data for the years directly affected by drought were omitted, the overall performance of the trees was significantly poorer than would have been predicted from models of beech growth. Some correlations were attempted with data for anthropogenic sulphur and nitrogen deposition. There was an indication of a negative association between growth reduction and levels of these pollutants but it was not statistically significant. There were no adequate data on ozone for use in analysis.

Sub-Top Dying of Conifers in the 1983/4 Dormant Season

Striking symptoms were observed in 1984 (Plate 6.4) on Sitka spruce, Norway spruce, Scots pine and other conifers at many locations along the west coast of Northern England and Scotland, (Fig. 6.5) (Redfern *et al.*, 1987b). Damage consisted primarily of shoot death which had occurred during the 1983/4 dormant season, foliar injury alone was rare. Dieback was not necessarily confined to the previous year's shoots but frequently extended into older internodes resulting in the death of fairly substantial branches. In the species most studied, Sitka spruce, the severity of damage increased both with elevation and tree age and was most severe in crops more than 45 years old at elevations greater than about 250 m. It aroused particular concern because of some similarity between the symptoms observed and those being reported at that time for Norway spruce in Germany. For example, although there was no needle yellowing, the dieback symptoms were concentrated in a zone just below the upper few whorls of the tree. Detailed studies showed that all damage had occurred during the 1983/4 dormant season and that it did not recur in the succeeding year. It was considered probable that a desiccation phenomenon was involved, this being linked to the rapid alternation that occurred between warm windy weather and cold spells (Redfern *et al.*, 1987b). Direct frost injury was not involved. Some possible interaction with pollution cannot be ruled out, although there was no association between the geographical distribution of the damage and any recorded high levels of air pollutants other than the concentration of pollutants due to the altitude effect.

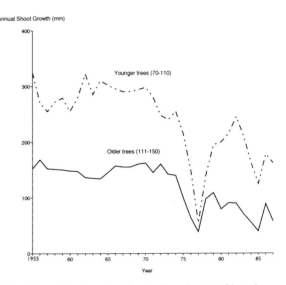

6.4 Growth trajectories for leading shoots of beech from 15 sites in southern England, are grouped for 2 age classes and show the dramatic effects of drought on growth in the subsequent year. (From Londsdale and Hickman, 1988.)

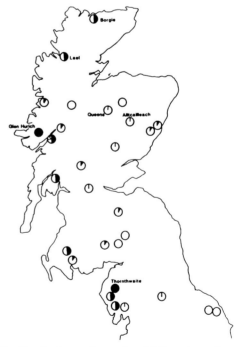

6.5 Distribution and severity of foliage browning and shoot death in Sitka spruce occurring during winter 1983–84 in plantations above 250 m. ○ – no damage; ● – severe damage. (From Redfern *et al.*, 1987b.)

7

Final Conclusions and Recommendations

Conclusions

Pollution climate

1. In the UK the majority of commercial forests are located in the north and west, often at considerable altitudes. Wet deposition of pollutants is large in these areas and concentrations of major ions in hill cloud have been shown to be very large relative to those in rain. It is clear that in these upland areas, tree foliage is exposed to very large concentrations of pollutants such as sulphate, nitrate, ammonium and acidity.

2. Inputs of sulphur and nitrogen to trees of southern and eastern England are dominated by dry deposition. Deposition of gaseous ammonia represents the largest atmospheric input of nitrogen to unfertilised vegetation in these areas. Concentrations of SO_2 and NO_2 in these areas are only moderate, but trees may also be exposed to frequent ozone episodes.

3. Amenity trees in urban areas are exposed to the largest concentrations of SO_2 and NO_2 in the UK. However **we know** that episodic exposure to ozone concentrations in cities is smaller than in the surrounding countryside due to ozone depletion by nitric oxide.

4. **We are certain** that forests increase the deposition of the pollutants nitric acid, ammonia, sulphur dioxide and of ions in cloud droplets, relative to short vegetation.

5. Although we understand the main mechanisms involved in dry deposition of sulphur dioxide to trees and can produce deposition maps, **we are uncertain** about the effect of gaseous ammonia on current sulphur dioxide

deposition estimates. At similar concentrations, the two gases may react on foliar surfaces and rates of sulphur dioxide deposition may thus be larger than modelled estimates. Further research is needed to improve our predictive capacity and to justify modification of estimates in critical load calculations.

6. Further investigations into deposition of nitrogen dioxide to vegetation are needed before confident estimates of nitrogen dioxide deposition to forests can be provided.

7. The frequency of ozone episodes and maximum concentrations observed in the north and west of Britain are lower than those further south.

8. There are between 50 and 200 hours per year when ozone concentrations exceed 60 ppb in southern England. The duration of exposure to these episodic concentrations is greatest in warm years.

Surveys of tree health

9. Although consistent surveys of crown condition in trees in the UK have been made since 1987, **we still cannot be certain** that tree health on a national basis is good or bad from the early studies. Following comparisons between methods used in the different countries, the most recent UNECE survey in 1991 shows tree health in the UK to be generally poorer than in other European countries.

10. **There is no evidence of** a widespread decline in crown condition or indeed in tree health in the UK. However **we conclude** from assessments of crown condition and other characters that a recent decline in condition of

beech has occurred in some areas of southern England. Results of the most recent UN-ECE survey also show a marked deterioration in the condition of broadleaved trees between 1990 and 1991.

11. Droughts have been associated with a deterioration in tree health particularly of beech and it is possible that the opening up of stands following deaths from the droughts of 1976, 1989 and 1990 and from storm damage has also contributed to the deterioration of the remaining trees.

12. The crown density of most species is less than expected for 'ideal' trees. **We are not certain** of the cause(s) of the density loss. We know from surveys that factors such as stand density and tree age can significantly affect crown condition. Insect and fungal attacks which also affect crown condition can be identified with some confidence, but at the present time the effects of pollutants on tree crowns cannot confidently be assessed from surveys.

13. It is difficult to separate the effects of pollutants from the effects of other environmental variables in survey data. Consequently, **we believe** that survey results alone cannot be used to determine the influence of atmospheric pollutants on tree health.

Effects of factors other than air pollution

14. Many instances of poor growth/tree damage in the UK are caused by factors other than air pollution.

15. **We know** that adverse climatic events can damage trees. Damage from high winds and unseasonal frosts occurs frequently in plantation forests and fruit orchards and soil water availability can limit growth in lowland and upland areas. Other soil conditions are also important. Nutrient deficiencies can restrict growth, often inducing a variety of foliar symptoms. Excesses of certain heavy metals can be toxic, causing poor or malformed growth.

16. **We also know** that conspicuous damage symptoms observed in crowns of mature trees are frequently caused by biotic agents, particularly fungi and insects.

Experimental evidence

17. We regard it as inevitable that much of the experimental evidence for effects of air pollutants on tree health will continue to be based mainly on relatively short term experiments on juvenile trees. However **we are not certain** that this evidence can be used to draw accurate conclusions about long term impacts of pollutant stress on mature trees on forest ecosystems.

18. Exclusion of gaseous pollutants in filtration experiments has improved the growth of beech and Norway spruce at a site in southern England but not at sites in the Pennines or in Scotland. The evidence suggests that ozone is the main pollutant responsible for the observed effect in southern England.

19. **We are certain** from fumigation experiments that exposure to ozone at current rural concentrations has reduced root growth of beech and at somewhat higher concentrations, has also reduced the above-ground growth of Sitka spruce.

20. Evidence suggests that the effect of ozone on biomass production may be cumulative, emphasising the importance of long term experiments.

21. **We have little evidence** that current rural concentrations of sulphur dioxide and nitrogen dioxide are directly reducing growth of major tree species in the UK.

22. **We have evidence** from experiments that acid mists (at concentrations within the range observed at upland locations) can reduce the growth of Sitka spruce. The observed reductions in stem volume of Sitka spruce occurred in the absence of visible injury symptoms.

23. In areas of high sulphur dioxide concentrations, co-deposition of sulphur dioxide and ammonia may result in increased inputs of sulphur and nitrogen to trees. **We are not certain** of the implications of this increase for tree health.

24. **We are certain** that realistic concentrations of gaseous pollutants and acid deposition can cause subtle changes in morphology, physiology and biochemistry of trees without obviously reducing growth. These changes may alter the sensitivity of trees to other environmental stresses. For example **we know** that:

    i) from experiments, ozone can increase the sensitivity of young trees to winter stress. However **we have no unequivocal evidence** that this is occurring under field conditions in the UK.

    ii) exposure to ozone can impair control of water loss in conifers. **We believe** that this effect may be long lasting, predisposing conifers to subsequent drought events or winter desiccation. But **we do not believe** that mature trees will be killed by this effect.

    iii) exposure to concentrations of sulphur dioxide and nitrogen dioxide found in the more polluted areas of the UK can increase the performance of certain pests such as green spruce aphid.

25. **Further information** is required to enable us to evaluate the impact of air pollutants on winter injury, water stress and disease severity.

26. On the basis of the experimental evidence and observed pollutant concentrations and deposition rates, **we conclude** that the air pollution climate in some areas of the UK may be detrimental to tree health.

Soils

27. Evidence suggests that increased acidification of forest soils in parts of the UK has been caused by deposition of pollutants.

28. **Our judgement is** that tree decline in which magnesium deficiency is implicated is unlikely in the UK because of large marine inputs of magnesium.

29. **We cannot yet demonstrate** that there are soil cation deficiencies induced by large inputs of ammonia in the UK. Further data on the spatial variations of atmospheric concentrations of ammonia are needed.

30. In a number of upland forest sites in UK, unexpectedly large amounts of nitrate are present in soil drainage and stream water. This may indicate a nutritional disorder of the crop. In central Europe such nitrate leaching has been associated with forest decline.

31. **We believe,** from limited data available, that acidification of soils is leading to a reduction in mineralisation and hence to a reduction in decomposition rates, with important implications for nutrient cycling in forests. Soil acidification can cause a shift in populations of root mycorrhizal fungi which are essential for nutrient uptake.

64

Case histories

32. Acute air pollution damage to trees is rare in Britain, but **we know** of instances where air pollution has been the primary cause of tree damage in areas close to pollution sources.

33. **We strongly suspect** that air pollution has caused damage to trees in areas remote from pollution sources. In such cases it is not always possible to prove the role of pollutants as causal agents.

34. **We believe** that instances exist where air pollution may be a **contributory** factor to tree damage, in association with other stresses. In many of these cases, **we cannot prove** that air pollution is involved but it must remain a suspect.

Recommendations

1. Inputs of atmospheric nitrogen in many areas of the UK are now known to be dominated by reduced nitrogen (ammonia). As the decision to reduce emissions of oxides of nitrogen has already been taken, some action on controlling ammonia emissions in the UK should be considered.

2. Inputs of gaseous ammonia probably represent the largest atmospheric nitrogen input to unfertilised vegetation over much of eastern and southern Britain. It is thus necessary to understand the relative importance of reduced nitrogen and oxidised nitrogen inputs throughout the UK and to study the ratio of NH_x/NO_x deposition. Further experimental work on the effects of gaseous ammonia on sulphur dioxide deposition estimates is also needed.

3. Forests increase the inputs of acidifying pollutants to soils and catchments. Forest soils have been acidified in parts of the UK, with implications for fresh water ecology and for the quality of drinking water supplies. Future land use policy will thus need to take into account the effects of afforestation on soils and water.

4. Plans for the creation of new 'urban forests' envisaged extensive planting in areas which may have relatively high concentrations of gaseous pollutants. The possible damaging effects of these pollutants will need to be considered as a factor in planning the location of these forests and in the choice of species.

5. The available evidence suggests that ozone is likely to adversely affect tree health in parts of Britain. Emission controls to reduce ozone should take into consideration the critical levels of ozone which affect tree health.

6. Current research and survey effort is primarily directed towards commercial forestry. The health of trees at sites of high conservation value should receive greater emphasis in future surveys and monitoring.

7. In the future development of tree and soil survey programmes it is important to liaise and co-operate with other European countries so that surveys can be more comparable and efforts to understand tree decline may be more effectively pooled.

8. It is clearly important to continue and improve monitoring of primary and secondary pollutants in the principal regions of UK forest production. Better estimates of exposure, dose and concentration of major pollutants at experimental forest sites are required.

9. Because of their structure, trees receive larger inputs of the pollutants nitric acid, ammonia, sulphur dioxide and cloud droplets than short vegetation. This has important implications for tree health and for pollutant loading. The deposition rates of pollutants

on forests should thus be investigated further so that deposition estimates can be provided for representative sites.

10. Investigations into cloud water composition and deposition should be extended to include altitude enhancement of deposition in upland areas. The effects of ozone episodes on tree health at high elevations also require further research.

11. A small number of sites should be established/developed, preferably with different pollution climates, for the purpose of carrying out intensive surveys of tree health and associated analyses of atmospheric and soil factors.

12. Surveys alone cannot conclusively demonstrate causal links between tree health and air pollution. It is thus vital that a programme of long-term experiments employing air filtration techniques and controlled exposure to realistic concentration of gases and mists is maintained. Some work must be done on trees beyond the seedling and sapling stages.

13. The uncertainty in extrapolating from effects of pollutants on young trees to effects of mature trees and ultimately forest stands has already been emphasised. It is recognised that it is not possible to conduct extensive experiments with mature trees, and thus future research should be directed towards determining the limitations of these extrapolations. Manipulation of forest environments may be useful in this context.

14. Research on interactions between pollutant, biotic and abiotic stresses should be continued, with emphasis again on long term experiments. Future studies should include various climate change scenarios and elevated concentrations of CO_2.

15. Work is needed on the impacts of high aluminium concentrations in soil solution on tree root growth and function in upland soils, and on the influence of high concentrations of ammonium and aluminium on the root uptake of other nutrients.

16. Dendrochronology (tree ring analysis) is already used in assessment of tree decline in other countries and should be introduced into UK research programmes. This should extend beyond measurements of growth to encompass changes in the chemical composition of tree rings.

17. There is a need to integrate research and modelling. Models should be capable of making useful predictions of risks to trees as improved knowledge of the complex links between climate, tree health and atmospheric pollution becomes available.

18. Where instances of tree damage occur, multidisciplinary analyses should be made. These analyses should draw as appropriate on the skill of specialists in fields such as pathology, soil science, pollution science, entomology, mycology, plant physiology and climatology.

References

Abrahamsen, G. & Stuanes, A.O. 1980. Effects of simulated rain on the effluent from lysimeters with acid, shallow soil, rich in organic matter. In: *Ecological Impact of Acid Precipitation – Proceedings of an International Conference*, edited by D. Drablos & A. Tollen. SNSF-Project, Norway

Adams, W.A., Ali, A.Y. & Lewis, P.J. 1990. Release of cationic aluminium from acidic soils into drainage waters and relationships with land use. *J. Soil Sci.* **41**, 255–268.

Aminu-Kano, M., McNeill, S. & Hails, R.S. 1991. Pollutant, plant and pest interactions: the grain aphid *Sitobion avenae*. *Agric. Ecosystem and Environ.*, **33**, 233–243.

Anon. 1989. *Waldzustandsbericht. Ergebnisse der Waldschadens-erhebung 1989.* Bundesministerium für Ernährung, Landwirtschaft und Forsten, Bonn.

Athati, S. & Kramer, H. 1989. Problematik der Zuwachsuntersuchungen in Buchenbeständen mit neuartigen Schadsymptomen. *Allgemeine Forst- und Jagdzeitung* **160**, 1–8.

Barbour, D.A. 1988. The pine looper in Britain and Europe. In: *Dynamics of Forest Insect Populations. Patterns, causes, implications*, edited by A.A. Berryman. New York, Plenum, 291–308.

Barnes, J.D. & Davison, A.W. 1988. The influence of ozone on the winter hardiness of Norway spruce. (*Picea abies*) (L.) Karst.). *New Phytol.*, **108**, 495–504.

Barnes, J.D., Davison, A.W. & Booth, T.A. 1988. Ozone accelerates structural degradation of epicuticular wax on Norway spruce needles. *New Phytol.*, **110**, 309–318.

Barnes, J.D., Eamus, D., Davison, A.W., Ro-Poulsen, H. & Mortensen, L. 1990. Persistent effects of ozone on needle water loss and wettability in Norway spruce. *Environ. Pollut.*, **63**, 345–363.

Bauch, J. 1983. Biological alterations in the stem and root of fir and spruce due to pollution influence. In: *Effects of Accumulation of Air Pollutants in Forest Ecosystems*, edited by B. Ulrich & J. Pankrath, 377–386. Dordrecht:Reidel Publ. Co.

von Becker, A. 1982. Aussaatversuch mit Bucheckern im Gewachshaus. Landesanst. f. Okologie Landschaftsentw. u. Forstpl. Nordrhein-Westfalen, 37–42.

Berden, M., Nilsson, S.I., Rosen, K. & Tyler, G. 1987. *Soil acidification – Extent, causes and consequences.* National Swedish Environmental Protection Board, Solna, Sweden.

Bewley, R.J.F. & Stotzky, G. 1983. Simulated acid rain (H_2SO_4) and microbial activity in soil. *Soil Biol, Biochem.* **15**, 425–429.

Billett, M.F., Fitzpatrick, E.A. & Cresser, M.S. 1988. Long-term changes in the acidity of forest soils in North-east Scotland. *Soil Use & Manage.,* **4**, 102–107.

Billett, M.F., Parker-Jones, F., Fitzpatrick, E.A. & Cresser, M.S. 1990. Forest soil chemical changes between 1949/50 and 1987. *J. Soil Sci.,* **41**, 133–145.

Binns, W.O., Mayhead, G.J. & Mackenzie, J.M. 1980. *Nutrient deficiencies of conifers in British forests.* Forestry Commisssion Leaflet 76. London: HMSO.

Binns, W.O., Redfern, D.B., Boswell, R.C. and Betts, A.J.A. 1986. *Forest health and air pollution: 1985 survey.* Forestry Commission Research and Development Paper 147, Forestry Commission, Edinburgh.

Binns, W.O., Redfern, D.B., Rennolls, K. and Betts, A.J.A. 1985. *Forest health and air pollution: 1984 survey.* Forestry Commission Research and Development Paper 142, Forestry Commission, Edinburgh.

Bolsinger, M. & Flückiger, W. 1989. Amino-acid changes by air pollution and aphid infestation. *Environ. Pollut.,* **56**, 209–216.

Bosshard, W. (ed.) 1986. *Kronenbilder.* Eidgenössische Anstalt für das forstliche Versuchswesen, Birmensdorf.

Bower, J.S., Lampert, J.E., Stevenson, K.J., Atkins, D.H.F. & Law, D.V. 1989. *The Results of a National Survey of Ambient Nitrogen Dioxide Levels in Urban Areas of the United Kingdom.* Warren Spring Laboratory.

Brasier, C.M. 1991. *Ophiostoma novo-ulmi* sp. nov., causative agent of current Dutch elm disease pandemics. *Mycopathologia,* **115**, 151-161.

Brasier, C.M. & Gibbs, J.N. 1973. Origin of the Dutch elm disease epidemic in Britain. *Nature, London,* **242**, 607–609.

Braun, S. & Flückiger, W. 1989. Effect of ambient ozone and acid mist on aphid development. *Environ. Pollut.,* **56**, 177–187.

Brown, K.A. 1987. Chemical effects of pH 3 sulphuric acid on a soil profile. *Wat. Air & Soil Pollut.,* **32**, 201–210.

Brown, K.A., Roberts, T.M. & Blank, L.W. 1987. Interaction between ozone and cold sensitivity in Norway spruce: a factor contributing to the forest decline in central Europe? *New Phytol.,* **105,** 149–155.

Brown, V.C. & Bell, J.N.B. 1990. The value of closed chambers experiments for studying biotic interactions with plants and air pollution. In: *Environmental Research with Plants in Closed Chambers,* edited by H.D. Payer, T. Pfirrmann and P. Mathy, 322–329. Air Pollution Research Report, No 26. CEC, Brussels.

van Breemen, N., Boderie, P.M.A. & Bootlink, H.W.G. 1989. Influence of airborne ammonium sulphate on soils of an oak woodland ecosystem in the Netherlands: seasonal dynamics of solute fluxes. In: *Acidic precipitation Volume 1 Case Studies.* edited by D.C. Adriano & M. Hava. New York: Springer Verlag.

Bucher, J.B., Schiller, G. & Siegwolf, R.T.W. 1988. Effects of ozone and/or water stress on xylem pressure, transpiration and leaf conductance in fir (*Abies alba* Mill.). In: *Relationships Between Above and Below Ground Influences of Air Pollutants on Forest Trees,* edited by J. Bervaes, P. Mathy & P. Evers, COST/CEC. Report No. 16, Brussels, Belgium.

Cannell, M.G.R. 1985. Analysis of risks of frost damage to forest trees in Britain. In: *Crop Physiology of Forest Trees,* edited by R.M.A. Tigerstedt, P. Puttonen, V. Koski, 152–166. University of Helsinki, Finland.

Cannell, M.G.R. & Smith, R.I. 1984. Spring frost damage on young *Picea sitchensis.* 2. Predicted dates of budburst and probability of frost damage. *Forestry,* **57,** 62–81.

Cape, J.N., Fowler, D., Eamus, D., Murray, M.B., Sheppard, L.J. & Leith, I.D. 1990. Effects of acid mist and ozone on frost hardiness of Norway spruce seedlings. In: *Environmental Research with Plants in Closed Chambers,* edited by H.D. Payer, T. Pfirrman & P. Mathy, 331–335. Air Pollution Research Report 26. CEC, Brussels.

Cape, J.N., Leith, I.D., Fowler, D., Murray, M.B., Sheppard, L.J., Eamus, D. & Wilson, R.H.F. 1991. Sulphate and ammonium in mist impair the frost hardening of red spruce seedlings. *New Phytol.,* **118,** 119–126.

Cape, J.N., Paterson, I.S., Wellburn, A.R., Wolfenden, J., Mehlhorn, H., Freer-Smith, P.H. and Fink, S. 1988. *Early diagnosis of forest decline – report of a one-year pilot study.* Institute of Terrestrial Ecology, Grange-over-Sands.

Carney, J.L., Garrett, U.E. & Hendrick, H.G. 1978. Influence of air pollutant gases on oxygen uptake of pine roots with selected ectomycorrhizae. *Phytopathology,* **68,** 1160–1163.

Carter, C.I. 1972. *Winter temperatures and survival of the green spruce aphid.* Forestry Commission Forest Record 84. HMSO, London.

Catt, J.A. 1985. Natural soil acidity. *Soil Use & Manage.*, **1**, 8–10.

Citrone, S.D. 1989. *The effects of sulphur dioxide and nitrogen dioxide on the relationship between Auchenoryncha species and their host plants.* Ph.D. Thesis, University of London.

Commission of the European Communities 1989. *European Community Forest Health Report 1987–1988.* Commission of the European Communities, Luxembourg.

Coutts, M.P. (Editor) 1993. *Decline of Sitka spruce in the South Wales coalfield.* Forestry Commission Bulletin (In preparation).

Darrall, N.M. 1989. The effect of air pollutants on physiological processes in plants. *Plant Cell Environ.*, **12**, 1–30.

Davidson, S.R., Ashmore, M.R. & Garretty, C. 1992. Effects of ozone and water deficit on the growth and physiology of *Fagus sylvatica*. *For. Ecol. Manage.*, **51**, 187–194.

Davison, A.W., Barnes, J.D. & Renner, C.J. 1988. Interactions between air pollutants and cold stress. In: *Air Pollution and Plant Metabolism,* edited by S. Schulte-Hostede, N.M. Darrell, L.W. Blank and A.R. Wellburn, 307–328. Elsevier, London.

Day, W.R. & Peace, T.R. 1946. *Spring frosts.* Forestry Commission Bulletin, **18**, 111 pp.

Deans, J.D. 1979. Fluctuations of the soil environment and fine root growth in a young Sitka spruce plantation. *Pl. Soil*, **52**, 195–208.

Denstorf, O., Heeschen, G. and Kenneweg, H. 1984. Ergebnisse der grossräumigen Inventuren von Waldschäden 1983 mit Farb-Infrarot-Luftbildern in südlichen Schleswig-Holstein. *Allgemeine Forst und Jagdzeitung* **155**, 126–131.

Diamandis, S. 1979a. Top-dying of Norway spruce, *Picea abies* with special reference to *Rhyzosphaera kalkhoffii* iv Top-dying of Norway spruce and climate. *Eur. J. For. Path.*, **9**, 77–88.

Diamandis, S. 1979b. Top-dying of Norway spruce, *Picea abies* with special reference to *Rhizosphaera kalkhoffii* vi Evidence related to the primary cause of Top-dying. *Eur. J. For. Path.*, **9**, 183–191.

Dighton, J. & Skeffington, R.A. 1987. Effects of artificial acid precipitation on the mycorrhizaes of Scots pine seedlings. *New Phytol.*, **107**, 191–202.

van Dijk, H.F.G., Creemers, R.C.M., Rijniers, J.P.L.W.M. & Roelofs, J.G.M. 1989. Impact of artificial ammonium-enriched rainwater on soils and young coniferous trees in a greenhouse. Part I: Effects on the soils. *Environ. Pollut.*, **62**, 317–336.

van Dijk, H.F.G. & Roelofs, J.G.M. 1988. Effects of excessive ammonium deposition on the nutritional status and condition of pine needles. *Physiol. Plant.* **73**, 494–501.

van Dijk, H.F.G., De Louw, M.H.J., Roelofs, J.G.M. & Verburgh, J.J. 1990. Impact of artificial, ammonium enriched rainwater on soils and young coniferous trees in a greenhouse. Part II: Effect on the trees. *Environ. Pollut.*, **63**, 41–59.

Dobson, M.C. 1991. *De-icing salt damage to trees and shrubs.* Forestry Commission Bulletin 101, 64 pp.

Dobson, M.C., Taylor, G. & Freer-Smith, P.H. 1990. The control of ozone uptake by *Picea abies* (L.) Karst and P. *sitchensis* (Bong.) Carr. during drought and interacting effects on shoot water relations. *New Phytol.*, **116**, 465–474.

Dohmen, C.P., McNeill, S. & Bell, J.N.B. 1984. Air pollution increases *Aphis fabae* pest potential. *Nature*, **307**, 52–53.

Dowding, P. 1988. Air pollutant effects on plant pathogens. In: *Air Pollution and Plant Metabolism*, edited by S. Schulte-Hostede, N.M. Darrall, L.W. Blank and A.R. Wellburn, 329–355. Elsevier, London.

Draaijers, G.P.J., Ivens, W.P.M.F., Bos, M.M. & Bleuten, W. 1989. The contribution of ammonia emissions from agriculture to the deposition of acidifying and eutrophying compounds onto forests. *Environ. Pollut.*, **60**, 55–66.

Durrant, D., Waddell, D.A., Houston, T. & Benham, S. 1991. *Air Quality and Tree Growth in Open-top Chambers.* Forestry Commission Research Information Note 20B.

Eamus, D., Barnes, J.D., Mortensen, L., Ro-Poulsen, H. & Davison, A.W. 1990. Persistent stimulation of CO_2 assimilation and stomatal conductance by summer ozone fumigation in Norway spruce. *Environ. Pollut.*, **63**, 365–379.

Eamus, D. & Fowler, D. 1990. Photosynthetic and stomatal conductance responses to acid mist of red spruce seedlings. *Plant Cell Environ.*, **13**, 349–357.

van der Eerden, L.J., Gremmen, M.H.M., van Dijk, C.J. & Steenbergen, P. 1989. *Luchtverontreiniging als mogelijke oorzaak van gewasschade in het Westland, (September 1989)*. Report from IPO (Instituut Voor Plantenziekten kundig Onderzoek).

Edlin, H.L. 1962. A modern sylva or a discourse of forest trees. *Q. J. of For.*, **56**, 196–205.

Farquhar, G.D., Firth, P.M., Wetselaar, R. & Wier, B. 1980. On the gaseous exchange of ammonia between leaves and the environment: determination of the ammonia compensation point. *Plant Physiol.*, **66**, 710–714.

Farrar, J.F., Relton, J. and Rutter, A.J. 1977. Sulphur dioxides and scarcity of *Pinus sylvestris* in the industrial Pennines. *Environ. Pollut.*, **14**, 63–68.

Fielding, N.J., Evans, H.F., Williams, J.M. & Evans, B. 1991. The distribution and spread of the great European spruce bark beetle, *Dendroctonus micans* in Britain – 1982–1989. *Forestry*, (In press).

Flückiger, W., Leonardi, S. & Braun, S. 1988. Air pollution effects on foliar leaching. In: *Scientific Basis of Forest Decline Symptomatology*, edited by J.N. Cape & P. Mathy, 160–169. COST/CEC, Brussels, Belgium.

Flückiger, W., Sury, R., Sinkemagel, C., Braun, S. & Hiltbrunner, E. 1990. The development of plant pathogens and pests in artificial environments. In: *Environmental Research in Closed Chambers*, edited by H.D. Payer, T. Pfirrman and P. Mathy, 341–346. Air Pollution Research Report 26. CEC, Brussels.

Forrest, G.I. 1980. Genotypic variation among native Scots pine populations in Scotland based on monoterpene analysis. *Forestry* **53**, 101–128.

Forrest, G.I. 1982. Relationship of some European Scots pine populations to native Scottish woodlands based on monoterpene analysis. *Forestry* **55**, 19–37.

Fowler, D. & Cape, J.N. 1982. Air pollutants in agriculture and horticulture. In: *Effects of Gaseous Air Pollution in Agriculture and Horticulture*, edited by M.H. Unsworth and D.P. Ormrod, 3–26. Butterworth Scientific.

Fowler, D., Cape, J.N., Nicholson, I.A., Kinnaird, J.W. & Paterson, I.S. 1980. The influence of polluted atmosphere on cuticle degradation in Scots pine. In: *Ecological impact of acid precipitation*, edited by D. Drablos & R. Tollan, 146–150. SNSF Project, Oslo.

Fowler, D., Cape, J.N. & Unsworth, M.H. 1989a. Deposition of atmospheric pollutants on forests. *Phil. Trans. R. Soc. Lond. B.*, **324**, 247–265.

Fowler, D. & Pitcairn, C. 1991. Regional foliar injury to trees and shrubs in Eastern England, September 1989. Internal Report, Institute of Terrestrial Ecology.

Fowler, D., Cape, J.N., Deans, J.D., Leith, I.D., Murray, M.B., Smith, R.I., Sheppard, L.J. & Unsworth, M.H. 1989b. Effects of acid mist on the frost hardiness of red spruce seedlings. *New Phytol.*, **113**, 321–335.

Freer-Smith, P.H. 1985. The influence of SO$_2$ and NO$_2$ on the growth, development and gas exchange of *Betula pendula* (Roth). *New Phytol.*, **99**, 417–430.

Freer-Smith, P.H. & Dobson, M.S. 1989. Ozone flux to *Picea sitchensis* (Bong.) Carr and *Picea abies* (L.) Karst, during short episodes and the effects of these on transpiration and photosynthesis. *Environ. Pollut.*, **59**, 161–176.

Freer-Smith, P.H. & Mansfield, T.A. 1987. The combined effects of low temperature and SO$_2$ + NO$_2$ pollution on the new season's growth and water relations of *Picea sitchensis. New Phytol.*, **106**, 225–237.

Gibbs, J.N. 1978. Development of the Dutch elm disease epidemic in southern England, 1971–6. *Annals app. Biol.*, **88**, 219–228.

Gibbs. J.N. & Greig, B.J.W. 1977. Some consequences of the 1975–1976 drought for Dutch elm disease in southern England. *Forestry*, **50**, 145–154.

Gibbs, J.N., Greig, B.J.W. & Hickman, I.T. 1987. An analysis of Peridermium stem rust of Scots pine in Thetford Forest in 1984 and 1985. *Forestry*, **60**, 203–218.

Gilbertson, P. & Bradshaw, A.D. 1990. The survival of newly planted trees in inner cities. *Abor. J.*, **15**, 287–390.

Gill, R.M.A. 1992a. A Review of damage by mammals in North Temperate Forests: 1. Deer. *Forestry*, **65**, 145–169.

Gill, R.M.A. 1992b. A Review of damage by mammals in North Temperate Forests: 2. Small Mammals. *Forestry*, **65**, 281–308.

Godwin, H. 1975. *History of the British flora. A factual basis for phytogeography.* 2nd edition. Cambridge University Press, Cambridge.

Green, F.H.W. 1964. A map of annual average potential water deficit in the British Isles. *J. appl. Ecol.*, **1**, 151–158.

Gregoire, J.C. 1988. The greater European spruce bark beetle. In: *Dynamics of Forest Insect Populations. Patterns, causes, implications,* edited by A.A. Berryman. New York: Plenum press, 455–478.

Greenhalgh, G.N. & Bevan, R.J. 1978. Response of *Rhytisma acerinum* to air pollution. *Trans. British Myc. Soc.*, **71**, 491–494.

Greig, B.J.W. 1984. Management of East England pine plantations affected by *Heterobasidion annosum* root rot. *Eur. J. For. Path.*, **14**, 392–397.

Greig, B.J.W. 1987. History of *Peridermium* stem rust of Scots pine in Thetford Forest, East Anglia. *Forestry*, **60**, 193–202.

Greig, B.J.W. & Gibbs, J.N. 1983. Control of Dutch elm disease in Britain. In: *Research on Dutch elm disease in Europe,* edited by D.A. Burdekin. Forestry Commission Bulletin 60, 10–16.

Greig, B.J.W., Gregory, S.F.C. & Strouts, R.G. 1991. *Honey fungus.* Forestry Commission Bulletin 100.

Gregory, S.C., MacAskill, G.A., Redfern, D.B. & Pratt, J.E. 1989. *Disease diagnostic service: Scotland and Northern England.* Annual Report on Forest Research 1989, **41.**

Gregory, S.C., Redfern, D.B., MacAskill, G.A. & Pratt, J.E. 1988. Advisory services: Scotland and Northern England. Report on Forest Research 1988. p. 39.

Grill, D., Pfeifhoffer, G., Halbwachs, G. & Waltingerm H. 1987. Investigations on epicuticular waxes of differently damaged spruce needles. *Eur. J. For. Pathol.,* **17,** 246–255.

Gruber, F. 1988. Der Fenstereffekt bei der Fichte. *Forst und Holz* **43,** 58–60.

Hallbacken, L. & Tamm, O. 1986. Changes in soil acidity from 1927 to 1982–1984 in a forest area of south-west Sweden. *Scand. J. For. Res.,* **1,** 219–232.

Hargreaves, K.J. & Atkins, D.H.F. 1988. The measurement of ammonia in the outdoor environment using passive diffusion tube samplers. AERE-R12568, United Kingdom Atomic Energy Authority, Harwell.

Hargreaves, K.J., Fowler, D., Storeton-West, R.L. & Duyzer, J.H. 1991. The exchange of nitric oxide, nitrogen dioxide and ozone between pasture and the atmosphere. *Environ. Pollut.,* **75,** 53–59.

Hauhs, M. 1989. Lange Bramke: An ecosystem study of a forested catchment. In: *Acidic Precipitation. Volume 1, Case Studies,* edited by D.C. Adriano and M. Havas. New York: Springer Verlag.

Hecht-Buchholz, C., Jorns, C.A. & Keil, P. 1987. Effect of excess aluminium and manganese on Norway spruce seedlings as related to magnesium nutrition. *J. Pl. Nutrition,* **10,** 1103–1110.

Henderson, D.M. and Faulkner, R. (eds.) 1987. Sitka spruce. *Proceedings of the Royal Society of Edinburgh, Section B* **93,** parts 1/2.

Heritage, S.G., Colins, S. & Evans, H.F. 1989. A survey of damage by *Hylobius abietis* and *Hylastes* spp. in Britain. In: *Insects affecting reforestation: biology and damage,* edited by R.I. Alfaro & S.G. Glover. Forestry, Canada, Victoria, British Columbia, 36–42.

74

Houlden, G., McNeill, S., Aminu-Kanu, M. & Bell, J.N.B. 1990. Air pollution and agricultural aphid pests. I. Fumigation experiments with SO_2 and NO_2. *Environ. Pollut.*, **67**, 305–314.

Houlden, G., McNeill, S., Craske, A. & Bell, J.N.B. 1991. Air pollution and agricultural aphid pests. II. Chamber filtration experiments. *Environ. Pollut.*, **72**, 45–55.

Hull, S.K. & Gibbs, J.N. 1991. *Ash dieback – a survey of non-woodland trees.* Forestry Commission Bulletin 93, 32 pp.

Hultberg, H., Ying-Hua Lee, Nystrom, U. & Nilsson, S.I. 1990. Chemical effects on surface-, ground- and soil-water of adding acid and neutral sulphate to catchments in southwest Sweden. In: *The Surface Waters Acidification Programme,* edited by B.J. Mason. Cambridge: Cambridge University Press.

Ineson, P. & Wookey, P.A. 1988. Effects of sulphur dioxide on forest litter decomposition and nutrient release. In: *Air Pollution and Ecosystems* edited by P. Mathy, D. 254–260. Reidel Publishing Company, Dordrecht.

Innes, J.L. 1988a. Forest health surveys: problems in assessing observer objectivity. *Can. J. For. Res.*, **18**, 560–565.

Innes, J.L. 1988b. Forest health surveys – a critique. *Environ. Pollut.*, **54**, 1–15.

Innes, J.L. 1990. Some problems with the interpretation of international assessments of forest damage. In: *Proceedings of the 19th World Congress of the International Union of Forestry Research Organisations, Montreal, August 5–11, 1990,* **2**, 380–387.

Innes, J.L. 1990. *Assessment of tree condition.* Forestry Commission Fieldbook 10.

Innes, J.L. and Boswell, R.C. 1987. *Forest health surveys 1987. Part 1: Results.* Forestry Commission Bulletin 74, HMSO, London.

Innes, J.L. and Boswell, R.C. 1989a. *Monitoring of forest condition in the United Kingdom 1988.* Forestry Commission Bulletin 88, HMSO, London.

Innes, J.L. and Boswell, R.C. 1989b. Sulphur contents of conifer needles in Great Britain. *Geo Journal* **19**, 63–66.

Innes, J.L. and Boswell, R.C. 1990a. *Monitoring of forest condition in Great Britain 1989.* Forestry Commission Bulletin 94, HMSO, London.

Innes, J.L. and Boswell, R.C. 1990b. Reliability, presentation and relationships amongst data from inventories of forest condition. *Can. J. For. Res.*, **20**, 790–799.

Innes, J.L., Boswell, R.C., Binns, W.O. and Redfern, D.B. 1986. *Forest health and air pollution: 1986 survey.* Forestry Commission Research and Development Paper 150, Forestry Commission, Edinburgh.

Jansen, A.E. & Dighton, J. 1990. *Effects of air pollution on ectomycorrhizaes.* Air pollution research report 25, CEC.

Jarvis, P.J. 1985. Transpiration and assimilation of tree and agricultural crops: The 'Omega factor'. In: *Attributes as Crop Plants,* edited by M.G.R. Cannell & J.E. Jackson. 460–480, I.T.E. Abbots Ripton, U.K.

Jarvis, P.J. & McNaughton, K.G. 1985. Stomatal control of transpiration scaling up from leaf to region. *Adv. Ecol. Res.,* **15,** 1–49.

Johannson, C. 1987. Pine forest – a negligible sink for atmospheric NO$_x$ in rural Sweden. *Tellus,* **39B,** 426–438.

Jurat, R. & Schaub, H. 1988. Effects of sulphur dioxide and ozone on ion uptake of spruce (*Picea abies* (L.) Karst) seedlings. *Z. pflanzenernahr, Budenk.,* **151,** 379–384.

Keane, K.D. & Manning, W.J. 1987. Effects of ozone and simulated acid rain and ozone and sulfur dioxide on mycorrhizal formation in paper birch and white pine. In: *Acid rain: Scientific and technical advances,* edited by R. Perry, R.M. Harrison, J.N.B. Bell & J.N. Lester, pp 608–613. London: Selper Ltd.

Keller, W. and Imhof, P. 1987. Zum Einfluss der Durchforstung auf die Waldschäden. 1. Teil: Erste Ergebnisse von Waldschadenuntersuchungen in Buchen-Durchforstungsflachen der EAFV. *Schweizerische Zeitschrift für Forstwesen* **138,** 39–54.

Keller, T. & Häsler, R. 1988. Some effects of long term fumigations with ozone on spruce (*Picea abies* (L.) Karst). *GeoJournal,* **17,** 277–278.

Kelly, J.M., Schaedle, M., Thornton, F.C. & Joslin, J.D. 1990. Sensitivity of tree seedlings to aluminium: II Red oak, Sugar Maple and European Beech. *J. environ. Qual.,* **19,** 172–179.

Kerstiens, G. & Lendzian, K.J. 1989. Interactions between ozone and plant cuticles II. Water permeability. *New Phytol.,* **112,** 21–27.

Kidd, N.A.C. & Thomas, M.B. 1988. The effects of acid mist on conifer aphids and their implication for tree health. In: *Air Pollution and Ecosystems,* edited by P. Mathy, 780–783. D. Reidel, Dordrecht.

Kinloch, B.B., Westfall, R.D. and Forrest, G.I. 1986. Caledonian Scots pine: origins and genetic structure. *New Phytol.,* **104,** 703–29.

Klumpp, G. & Guderian, R. 1989. Wirkung verschiedener Kombination von O$_3$, SO$_2$ und NO$_2$ auf Photosynthese und Atmung. *Staub,* **49,** 255–260.

Krause, G.H.M., Prinz, B. and Jung, K.D. 1983. Forest effects in West Germany. In: *Air pollution and the productivity of the forest*, 297–331. Pennsylvania State University.

Landmann, G. 1990. In: *La Santé Des Forêts 1990.* Ministère de l'agriculture et de la forêt, Paris.

Lee, H.S.J., Willson, A., Benham, S.E., Darrant, D.W.H., Houston, T. & Waddell, D.A. 1990a. *The effect of air quality on tree growth.* Research Information Note 182, Forest Research Station, Alice Holt Lodge, Wrecclesham, Surrey. GU10 4LH.

Lee, H.S.J., Willson, A., Benham, S.E., Darrant, D.W.H., Houston, T. & Waddell, D.A. 1990b. *The effect of air quality on the timing of tree shoot development.* Research Information Note 183, Forest Research Station, *op cit.*

Lee, J.J. & Weber, D.E. 1982. Effects of sulphuric acid irrigation on major cation and sulphate concentrations on water percolating through two model hardwood forests. *J. environ. Qual.,* **11,** 57–64.

Leith, I.D. & Fowler, D. 1987. Urban distribution of *Rhytisma acerinum* (Pers.) Fries (tar spot) on sycamore. *New Phytol.,* **108,** 175–181.

Lendzian, K.J., Nakajiima, A. & Ziegler, H. 1986. Isolation of cuticular membranes from various conifer needles. *Trees,* **1,** 47–53.

Levin, M.J. 1985. Ergebnisse und waldbauliche Konsequenzen einer Immissionsschadenserhebung im Privatforstbetrieb Sachsenwald. *Wald-hygiene* **16,** 19–22.

Like, D.E. & Klein, R.M. 1985. The effect of simulated acid rain on nitrate and ammonium production in soils from three ecosystems of Camels Hump mountain. *Soil Sci.,* **140,** 352–355.

Lines, R. 1987. *Choice of seed origins for the main forest species in Britain.* Forestry Commission Bulletin 66, HMSO, London.

Locke, G.M.L. 1987. *Census of woodlands and trees 1979–82.* Forestry Commission Bulletin 63, HMSO, London.

Lonsdale, D. 1986a. *Beech health study 1985.* Forestry Commission Research and Development Paper 146, Forestry Commission, Edinburgh.

Lonsdale, D. 1986b. *Beech health study 1986.* Forestry Commission Research and Development Paper 149, Forestry Commission, Edinburgh.

Lonsdale. D. 1989. Pruning practice. In: *Urban Forestry Practice,* edited by B.G. Hibberd. Forestry Commission Handbook 5, 94–100.

Lonsdale, D. & Wainhouse, D. 1987. *Beech bark disease.* Forestry Commission Bulletin 69, 24 pp.

Lonsdale, D. & Hickman, I.T. 1988. *Beech health study.* Report on Forest Research, 1988, 42–44. Edinburgh.

Lonsdale, D., Hickman, I.T., Mobbs, I.D. & Matthews, R.W. 1989. A quantitative analysis of beech health and pollution across southern Britain. *Naturwissenschaften,* 76, 571–573.

Lorenc-Plucinska, G. 1986. Effect of sulphur dioxide on the partitioning of assimilates in Scots pine (*Pinus sylvestris* L.) seedlings of different susceptibility to this gas. *Eur. J. For. Pathol.,* 16, 266–273.

Lucas, P.W., Cottam, D.A., Sheppard, L.J. & Francis, B.J. 1988. Growth responses and delayed winter hardening in Sitka spruce following summer exposure to ozone. *New Phytol.,* 108, 495–504.

Mahoney, M.J., Chevone, B.I., Stelly, J.M. & Moore, L.D. 1985. Influence of mycorrhizae on the growth of loblolly pine seedlings exposed to ozone and sulphur dioxide. *Phytopathology,* 75, 679–682.

Mansfield, T.A. 1988. Factors determining root-shoot partitioning. In: *Scientific Basis of Forest Decline Symptomatology,* edited by J.N. Cape & P. Mathy, 171–180. COST/CEC, Brussels, Belgium.

Mansfield, P.J., Bell, J.N.B., McLeod, A.R. & Wheeler, B.J. 1991. Effects of sulphur dioxide on the development of fungal diseases of winter barley in an open-air fumigation system. *Agric. Ecosystems & Environ.,* 33, 215–232.

Matzner, E., Khanna, P.K., Meiwes, K.J. & Ulrich, B. 1983. Effects of fertilization on the fluxes of chemical elements through different forest ecosystems. *Pl. & Soil,* 74, 343–358.

Matzner, E. & Ulrich, B. 1985. "Waldsterber": Our dying forests – Part II. Implications of the chemical soil condition for forest decline. *Experimentia,* 41, 578–584.

McLaughlin, S.B. 1985. Effects of air pollution on forests: a critical review. *J. Air Poll. Control Assoc.,* 35, 516–534.

McLeod, A.R., Holland, M.R., Shaw, P.J.A., Sutherland, P.M., Darrall, N.M. & Skeffington, R.A. 1990. Enhancement of nitrogen deposition to forest trees exposed to SO_2. *Nature,* 347, 277–279.

McNeill, Aminu-Kano, M., Houlden, G., Bullock, J.M., Citrone, S. & Bell, J.N.B. 1987. The interactions between air pollution and sucking insects. In: *Acid Rain: Scientific and Technical Advances,* edited by R. Perry *et al.,* 602–607. Selper Ltd., London.

McNeill, S. & Whittaker, J.B. 1990. Air pollution and tree-dwelling aphids. In: *Population Dynamics of Forest Insects* edited by A.D. Watt, S.R. Leather, M.D. Hunter and N.A.C. Kidd, 195–208. Intercept Ltd., Andover.

Mengel, K., Hogrebe, A.M.R. & Esch, A. 1989. Effect of acidic fog on needle surface and water relations of *Picea abies*. *Physiol. Plant*, **75**, 201–207.

Mohr, H. 1986. Die Erforschung der neuartigen Waldschaden eine Zwischenbilanz. *Biol. unserer Zeit*, **16**, 83–89.

Monteith, J.L. & Unsworth, M.H. 1990. *Principles of Environmental Physics*. Edward Arnold London.

Morrison, I.K. 1983. Composition of percolate from reconstructed profiles of two Jack Pine forest soils as influenced by acid input. In: *Accumulation of Air Pollutants in Forest Ecosystems*, edited by B. Ulrich & J. Pankrath. Dordrecht: Reidel.

Muir, P.S. and Armentano, T.V. 1988. Evaluating oxidant injury to foliage of *Pinus ponderosa:* a comparison of methods. *Can. J. For. Res.* **18**, 498–505.

Neighbour, E.A., Cottam, D.A. & Mansfield, T.A. 1988. Effects of sulphur dioxide and nitrogen dioxide on the control of water loss by birch (*Betula*) spp. *New Phytol.*, **108**, 149–157.

Nelson, D.G. 1990. Restocking with Sitka spruce on uncultivated gley soils – the effects of fencing, weeding and initial plant size on survival and growth. *Scottish Forestry*, **44**, 266–272.

Neufeld, M.S., Jernstedt, J.A. & Haines, G.L. 1985. Direct foliar effects of acid rain. *New Phytol.*, **99**, 389–406.

Neumann, M. 1989. Einfluss von Standortsfaktoren auf den Kronenzustand. In: *Forest decline and air pollution*, edited by J.B. Bucher and I. Bucher-Wallin, 209–214. Eidgenössische Anstalt für das forstliche Versuchswesen, Birmensdorf.

Norby, R.J., Taylor, G.E., McLaughlin, S.B. & Gunderson, C.A. 1986. Drought severity of red spruce seedlings affected by precipitation chemistry. Proc. Ninth North American Forest Biology Workshop, Stillwater, Oklahoma.

Norby, R.J., Weerasuriya, Y. & Hanson, P.J. 1989. Induction of nitrate reductase activity in red spruce needles by nitrogen dioxide and nitric acid vapour. *Can. J. For. Res.*, **19**, 889–896.

Paganelli, D.J., Seiler, J.R. & Feret, P.P. 1987. Root regeneration as an indicator of aluminium toxicity in loblolly pine. *Pl. Soil*, **102**, 115–118.

Pawsey, R.G. 1983. *Ash dieback survey: summer 1983.* Commonwealth Forestry Institute Occasional Paper No. 24, Oxford, 22 pp.

Philips, D.H. & Burdekin, D.A. 1982. *Diseases of forest and ornamental trees.* The MacMillan Press Ltd., London and Basingstoke.

Port, G.R. & Thompson, J.R. 1980. Outbreaks of insect herbivores on plants alongside motorways in the United Kingdom. *J. appl. Ecology.,* **17,** 649–656.

Potter, C.J. 1989. Establishment and early maintenance. In: *Urban Forestry Practice,* edited by B.G. Hibberd, Forestry Commission Handbook 5. 78–90.

Power, S.A. & Ashmore, M.R. 1991. *Beech Health and Air Pollution.* Final contract report to Nature Conservancy Council and Forestry Commission. Imperial College Centre for Environmental Technology, London, SW7 2PE.

Power, S., Ling, K. and Ashmore, M. 1989. *Native trees and air pollution.* Final report for contract HF3-03-326. Nature Conservancy Council, Peterborough.

Prinz, B. and Krause, G.H.M. 1989. State of scientific discussion about the causes of the novel forest decline in the Federal Republic of Germany. In: *Forest decline and air pollution,* edited by J.B. Bucher and I. Bucher-Wallin, 27–34. Eidgenössische Anstalt für das forstliche Versuchswesen, Birmensdorf.

Prinz, B., Krause, G.H.M. & Jung, K-D. 1987. Development and causes of novel forest decline in Germany. In: *Effects of Atmospheric Pollutants on Forests, Wetlands and Agricultural Ecosystems,* edited by T.C. Hutchinson & K.M. Meema, 1–24, Springer Verlag, Berlin.

Quine, C.P. 1991. Recent storm damage to trees and woodlands in southern Britain. In: *Research for Practical Arboriculture,* edited by S.J. Hodge, Forestry Commission Bulletin 97, 83–89.

Raynal, D.J., Joslin, J.D., Thornton, F.C., Schaedle, M. & Henderson, G.S. 1990. Sensitivity of tree seedlings to aluminium: III Red spruce and Loblolly pine. *J. environ. Qual.,* **19,** 180–187.

Redfern, D.B. 1982. *Spring frost damage on Sitka spruce.* Report on Forest Research 1982, 27.

Redfern, D.B. & Cannell, M.G.R. 1982. Needle damage in Sitka spruce caused by early autumn frosts. *Forestry* **55,** 39–45.

Redfern, D.B. & Gregory, S.C. 1991. *Needle browning and dieback of Scots pine in Northern Britain.* Report on Forest Research 1990, Edinburgh, 50–51.

Redfern, D.B., Gregory, S.C., MacAskill, G.A. & Pratt, J.E. 1987a. Pathology Advisory Service: Scotland and Northern England. Report on Forest Research 1987, Edinburgh, 42–43.

Redfern, D.B., Gregory, S.C., Pratt, J.E. & MacAskill, G.A. 1987b. Foliage browning and shoot death in Sitka spruce and other conifers in northern Britain during winter 1983–84. *Eur. J. For. Path.,* **17,** 166–180.

Redfern, D.B. & Rose, D.R. 1984. *Winter cold damage to pines.* Report on Forest Research 1984, Edinburgh, 33–34.

Rehfuess, K.E. 1987. Perceptions of forest diseases in central Europe. *Forestry,* **60,** 1–11.

Reich, P.B. & Lassoie, J.P. 1984. Effects of low level ozone exposure on leaf diffusive conductance and water use effeciency in hybrid poplar. *Plant Cell Environ.,* **7,** 661–668.

Reimer, J. & Whittaker, J.B. 1989. Air pollution and insect herbivores: observed interactions and possible mechanisms. *Insect-Plant Interactions,* **1,** 73–105.

RGAR 1990. *Acid Deposition in the United Kingdom, 1986–88.* Third report of the United Kingdom Review Group on Acid Rain. Department of the Environment Publications Sales Unit, South Ruislip.

Riding, R.T. & Percy, K.E. 1985. Effects of SO_2 and other air pollutants on the morphology of epicuticular waxes on needles of *Pinus strobus* and *Pinus banksiana. New Phytol.,* **99,** 555–563.

Roberts, T.M., Skeffington, R.A. and Blank, L.W. 1989. Causes of Type I spruce decline in Europe. *Forestry* **62,** 179–222.

Roloff, A. 1985a. Schadstufen bei der Buche. *Der Forst- und Holzwirt* **40,** 131–134.

Roloff, A. 1985b. Untersuchungen zum vorzeitigen Laubfall und zur Diagnose von Trockenschäden in Buchenbeständen. *Allgemeine Forst Zeitschrift* **40,** 157–160.

Rose, C. and Neville, M. 1985. *Tree dieback survey. Final Report.* Friends of the Earth, London.

Rose, D.R., 1991. *Ramichloridium* dieback of lodgepole pine. Report for Forest Research 1990, 50–52.

Rost-Siebert, K. 1983. Aluminium-Toxizitat und-Toleranz an Keimpflanzen von Fichte (*Picea abies Karst.*) und Buche (*Fagus silvatica* L.). *Allg. Forst. Holwirtsch.* Ztg. **26/27,** 680–686.

Ryan, P.J., Gessel, S.P. & Zososki, R.J. 1986. Acid tolerance of Pacific Northwest conifers in solution culture II. Effect of varying aluminium concentration at constant pH. *Pl. Soil,* **96,** 259–272.

Saunders, P.J.W. 1966. The toxicity of sulphur dioxide to *Diplocarpon rosae* Wolf causing blackspot of roses. *Ann. appl. Biol.*, **58**, 103–114.

Schöpfer, W. 1985. Das Schulungs- und Kontrollsystem der terristrischen Waldschadensinventuren. *Allgemeine Forst Zeitschrift* **40**, 1353–1357.

Schröder, H. and Aldinger, E. 1985. Beurteilung des Gesundheitzustandes von Fichte und Tanne nach der Benadelungsdichte. *Allgemeine Forst Zeitschrift* **40**, 1353–1357.

Schröder, W.H., Bauch, J. & Endeward, R. 1988. Microbeam analysis of Ca exchange and uptake in fine roots of spruce: influence of pH and aluminium. *Trees: Structure and Function*, **2**, 96–103.

Schulze, E.D., Hantschel, R., Werk, K.S. & Horn, R. 1987. Water relations of two Norway spruce stands at different stages of decline. In: *Forest Decline and Air Pollution,* edited by E.-D. Schulze, O.L. Lange & R. Oren, 341–351, Ecological Studies Vol. 77, Springer-Verlag, Berlin, Heidelberg.

Sensor, M. & Payer, H.D. 1989. The frost hardiness of young conifers exposed to different loads of ozone, sulphur dioxide in environmental chambers. In: *Air Pollution and Forest Decline,* edited by J.B. Bucher & I. Bucher-Wallin, 523–526. EAFV. Birmensdorf, Switzerland.

Skärby, L., Troeng, E. & Bostrom, C-A. 1987. Ozone uptake and effects on transpiration net photosynthesis and dark respiration in Scots pine. *Forest Sci.,* **33**, 801–808.

Speight, M.R. & Wainhouse D. 1989. *Ecology and Management of Forest Insects.* Clarendon Press, Oxford.

Stevens, P.A. & Hornung, M. 1988. Nitrate leaching from a felled Sitka spruce plantation in Beddgelert Forest, North Wales. *Soil Use & Manage.,* **4**, 3–9.

Stevens, P.A., Hornung, M. & Hughes, S. 1989. Solute concentrations, fluxes and major nutrient cycles in a mature Sitka spruce plantation in Beddgelert Forest, North Wales. *Forest Ecology & Management,* **27**, 1–20.

Stienen, H. & Bauch, J. 1988. Element content in tissues of spruce seedlings from hydroponic cultures simulating acidification and deacidification. *Pl. Soil,* **106**, 231–238.

Stroo, H.F. & Alexander, M. 1986. Available nitrogen and cycling in forest soils exposed to simulated acid rain. *Soil Sci. Soc. Am. J.,* **50**, 110–114.

Strouts, R.G. & Patch, D. 1983. *The cold winter of 1981–1982.* Report on Forest Research 1983, Edinburgh, 35–36.

Sucoff, E., Thornton, F.C. & Joslin, J.D. 1990. Sensitivity of tree seedlings to aluminium: I. Honeylocust. *J. environ. Qual.,* **19**, 163–171.

Sutton, M.A. 1990. *The surface/atmosphere exchange of ammonia.* Ph.D. Thesis, University of Edinburgh.

Taylor, C.M.A. 1986. *Forest fertilization in Great Britain.* Paper read before The Fertilizer Society of London, 11 December 1986.

Taylor, C.M.A. & Tabrush, P.M. 1990. *Nitrogen deficiency in Sitka spruce plantation.* Forestry Commission Bulletin 89. London: HMSO.

Taylor, G.E., Norby, R.J., McLaughlin, S.B., Johnson, A.H. & Turner, R.S. 1986. CO$_2$ assimilation and growth of red spruce seedlings in response to ozone, precipitation chemistry and soil type. *Oecologia,* **70,** 163–171.

Taylor, G. & Davies, W.J. 1990. Root growth of *Fagus sylvatica:* impact of air quality and drought at a site in southern Britain. *New Phytol.,* **116,** 457–464.

Taylor, G. & Dobson, M.C. 1989. Photosynthetic characteristics, stomatal responses and water relations of *Fagus sylvatica:* impact of air quality and drought at a site in southern Britain. *New Phytol.,* **113,** 265–273.

Taylor, G., Dobson, M.C., Freer-Smith, P.H. & Davies, W.J. 1989. *Tree physiology and air pollution in southern Britain.* Research Information Note **145,** Forestry Commission, UK.

Thiebaut, B. 1988. Tree growth, morphology and architecture, the case of beech: *Fagus sylvatica* L. In: *Scientific basis of forest decline symptomatology,* edited by J.N. Cape and P. Mathy, 49–72. Commission of the European Communities Air Pollution Report 15, Brussels.

Tickle, A. 1988. *Tree survey of Southern England.* Greenpeace Environmental Trust, Air Pollution Report 4, London.

Tiedemann, A.V., Ostlander, P., Firsching, K.H. & Fehrmann, H. 1990. Ozone episodes in southern lower Saxony (FRG) and their impact on the susceptibility of cereals to fungal pathogens. *Environ. Pollut.,* **67,** 43–59.

Turunen, M. & Huttunen, S. 1990. A review of the response of epicuticular wax of conifer needles to air pollution. *J. environ. Qual.,* **19,** 35–45.

Ulrich, B. & Matzner, E. 1986. Anthropogenic and natural acidification in terrestrial ecosystems. *Experimentia,* **42,** 344–350.

Ulrich, B., Mayer, R. & Khann, P.K. 1980. Chemical changes due to acid precipitation in a loess-derived soil in central Europe. *Soil Sci.,* **130,** 193–199.

UK TERG 1988. *The effects of acid deposition on the terrestrial environment in the United Kingdom.* HMSO, London.

UN-ECE 1988. *Forest damage and air pollution. Report of the 1987 forest damage survey in Europe.* United Nations, Convention on Long-Range Transboundary Air Pollution, Geneva.

UN-ECE 1990. *Forest damage and air pollution. Report of the 1988 forest damage survey in Europe.* United nations, Convention on Long-Range Transboundary Air Pollution, Geneva.

UN-ECE, 1992 *Forest Conditions in Europe,* Executive Summary 1992, Report from Programme Co-ordinating Centre – West Hamburg, Germany.

Wang, C. & Cootes, D.R. 1981. *Sensitivity classification of agricultural land to long term acid precipitation in eastern Canada.* (L.R.R.I. contribution 98). Agriculture Canada Research Board.

Warrington, S. 1987. Relationship between SO$_2$ dose and the growth of the pea aphid, *Acyrthosiphon pisum,* on peas. *Environ. Pollut.,* **43,** 155–162.

Warrington, S., Cottam, D.A. & Whittaker, J.B. 1989. Effects of insect damage on photosynthesis, transpiration and SO$_2$ uptake in sycamore. *Oecologia,* **80,** 136–139.

Warrington, S. & Whittaker, J.B. 1990. Interactions between Sitka spruce, the green spruce aphid, sulphur dioxide pollution and drought. *Environ. Pollut.,* **65,** 363–370.

Westman, L. 1989. A system for regional inventory of damage to birch. *International Congress on Forest Decline Research: State of Knowledge and Perspectives,* Poster Abstracts Volume 1, 57–58.

Westman, L. and Lesinski, J.A. 1986. *Kronutglesning och andra förändringar i grankronan.* Naturvardsverket Rapport 3262.

Willson, A., Waddell, D.A. & Durrant, D.W.H. 1987. *Experimental work on air pollution.* Research Information Note 121, Forest Research Station, *op cit.*

Wilson, M.J. 1989. Vulnerable soils and their distribution. In: *Acidification in Scotland 1988.* Edinburgh: Scottish Development Department.

Wookey, P.A. & Ineson, P. 1991. Chemical changes in decomposing forest litter in response to atmospheric sulphur dioxide. *J. Soil Sci.,* **42,** 615–628.

Worrell, R. 1987. *Predicting the Productivity of Sitka spruce on Upland Sites in Northern Britain.* Forestry Commission Bulletin, 72, HMSO London.

Zottl, H.W. 1990. Remarks on the effects of nitrogen deposition to forest ecosystems. *Pl. Soil,* **128,** 83–89.

Annex 1

Acknowledgements

The review group wishes to thank the many individuals and organisations who have contributed to this report, in particular Dr JD Barnes, Dr D Beagley, Dr KR Briffa, Dr KA Brown, Dr JN Cape, Professor MS Cresser, Dr M Coutts, Dr WJ Davies, Dr D Durrant, Dr HF Evans, Dr PH Freer-Smith, Dr JG Irwin, Mr CP Quine, Dr DB Redfern, Dr RA Skeffington, Dr G Taylor, Dr D Wainhouse, Professor A R Wellburn.

Annex 2

Membership of Terrestrial Effects Review Group

Professor MH Unsworth (Chairman)
Department of Physiology and
   Environmental Science
School of Agriculture
University of Nottingham
Sutton Bonington
Loughborough
LE12 5RD

Dr MR Ashmore
Centre for Environmental Technology
Imperial College of Science,
   Technology and Medicine
48 Prince's Gardens
LONDON
SW7 2PE

Dr MGR Cannell
Institute of Terrestrial Ecology
Bush Estate
Penicuik
Midlothian
EH6 OQB

Professor D Fowler
Institute of Terrestrial Ecology
Bush Estate
Penicuik
Midlothian
EH6 OQB

Dr JN Gibbs
Forestry Commission
Alice Holt Lodge
Wrecclesham
Farnham
Surrey
GU10 4LH

Professor M Hornung
Institute of Terrestrial Ecology
Merlewood Research Station
Windermere Road
Grange-over-Sands
Cumbria
LA11 6JU

Dr P Ineson
Institute of Terrestrial Ecology
Merlewood Research Station
Windermere Road
Grange-over-Sands
Cumbria
LA11 6JU

Dr JL Innes
Forestry Commission
Alice Holt Lodge
Wrecclesham
Farnham
Surrey
GU0 4LH

Dr PW Lucas
Division of Biological Sciences
Institute of Environmental and
   Biological Sciences
Lancaster University
LA1 4YQ

Professor TA Mansfield FRS
Division of Biological Sciences
Institute of Environmental and
   Biological Sciences
Lancaster University
LA1 4YQ

Mr RB Wilson (Executive Secretary)
Department of the Environment
Romney House
43 Marsham Street
LONDON
SW1P 3PY

Dr CER Pitcairn (Editorial Production)
Institute of Terrestrial Ecology
Bush Estate
Penicuik
Midlothian
EH26 OQB

Annex 3

Reports Prepared at the Request of the Department of the Environment

Acid Deposition in the United Kingdom
UK Review Group on Acid Rain, December 1983

Acid Deposition in the United Kingdom 1981–1985
UK Review Group on Acid Rain, August 1987
Copies of both are available from:
Warren Spring Laboratory
Gunnels Wood Road
Stevenage
Herts
SG1 2BX
Price £10.00 each.

Acidity in United Kingdom Fresh Waters
UK Acid Waters Review Group, April 1986.

Ozone in the United Kingdom
UK Photochemical Oxidants Review Group, February 1987.

Acid Deposition in the United Kingdom 1986–1988
UK Review Group on Acid Rain, September 1990.

Oxides of Nitrogen in the United Kingdom
Second Report on the UK Photochemical Oxidants Review Group, June 1990.
Copies available from:
DoE/DTP Publications Sales Unit
Building 1
Victoria Road
South Ruislip
Middlesex
HA4 ONZ
Price £10.00 each.

OXFORDSHIRE
COUNTY
LIBRARIES

Stratospheric Ozone
UK Stratospheric Ozone Review Group, August 1987.
Copies available from:
HMSO
(ISBN 0 11 7520187)
Price £9.95 each.

Stratospheric Ozone 1990
UK Stratospheric Ozone Review Group, June 1990.
Copies available from:
HMSO
(ISBN 0 11 752320 8)
Price £5.30 each.

The Effects of Acid Deposition on Buildings and Building Materials in the UK
Building Effects Review Group, 1989
Copies available from:
HMSO
(ISBN 0 11 7521787)
Price £11.25 each.

The Effects of Acid Deposition on the Terrestrial Environment in the United Kingdom
UK Terrestrial Effects Review Group, 1988.
Copies available from:
HMSO
(ISBN 0 11 752029 2)
Price £11.95 each.

The Potential Effects of Climate Change in the United Kingdom
UK Climate Change Impacts Review Group, 1991.
Copies available from:
HMSO
(ISBN 0 11 752359 3)
Price £8.50 each.

Urban Air Quality in the United Kingdom
Quality and Urban Air Review Group, Jan 1993.
Copies available from:
Department of Environment
PO Box 135
Bradford
West Yorkshire
BD9 4HN

Printed in the United Kingdom for HMSO
Dd 94822 c10   4/93